Function Approach to Transportation Projects

A Value Engineering Guide

Function Approach to Transportation Projects

A Value Engineering Guide

Muthiah Kasi

IUNIVERSE, INC.
NEW YORK   BLOOMINGTON

Function Approach to Transportation Projects - A Value Engineering Guide

iUniverse books may be ordered through booksellers or by contacting:

iUniverse
1663 Liberty Drive
Bloomington, IN 47403
www.iuniverse.com
1-800-Authors (1-800-288-4677)

Because of the dynamic nature of the Internet, any Web addresses or links contained in this book may have changed since publication and may no longer be valid.

ISBN: 978-1-4401-5144-6 (sc)
ISBN: 978-1-4401-5145-3 (dj)
ISBN: 978-1-4401-5146-0 (ebk)

Printed in the United States of America

iUniverse rev. date: 6/16/2009

Preface

This book describes value engineering (VE), a logical, structured method of solving problems within a team-oriented approach. Larry Miles, widely known as the father of value engineering, introduced this concept during World War II. His subsequent practice of this technique had a strong emphasis on creativity. This book uses a specific transportation project as a case study to demonstrate value engineering applications.

Value engineering is practiced with a team approach, usually three to six people. However, after learning these techniques, you will find that VE can be performed by an individual on small elements.

The Details of the Content

The transportation case study is from an actual Benesch project. The VE methodology applied during this study is described in sequential order, so the reader can gain an understanding of how the different components relate to each other. Each chapter uses this case study as the primary example to demonstrate the theory discussed. In addition, other examples are used to supplement the reader's understanding of the subject discussed in the chapter.

Chapter 1 introduces the VE Job Plan. It describes the phases, with questions and rules for each phase. The VE Job Plan includes the Information Phase, Speculation Phase, Evaluation Phase, Development Phase and Presentation Phase.

Chapters 2 through **6** describe the Information Phase. The VE methodology stresses the importance of understanding and analyzing the problems. The majority of the VE effort, as much as 60% , is used during the Information Phase.

Chapter 2 defines the people behind the problem or project. Identifying the stakeholders and understanding their constraints, needs and desires are the steps learned here. In particular, recognizing stakeholders roadblocks, in the form of constraints, is key to the implementation of any new idea.

Chapter 3 describes the unique VE approach to functions. Using the conflicting needs and desires of various stakeholders of the project, the VE process expresses them in terms of project functions. Functions are the reason the final project will be accepted as a solution.

Chapter 4 demonstrates a way to organize functions in a structured manner. Function Analysis System Techniques (FAST) separates VE from other methodologies. In this book, it is described as a "Function Logic Diagram" instead of a "FAST diagram." Either name is acceptable.

Chapter 5 stresses that all decisions are based on cost or affordability. Cost of the project can be determined at various levels of the VE, depending upon the details that are available at that time. The concept of elemental cost, as opposed to the trade cost of material, is needed to perform function analysis.

Chapter 6 describes the final step of the Information Phase. Function analysis distributes the cost of the project to each function. It also questions the worth of the function cost. This step will reveal the valuable functions as well as the mismatched functions. It sets the direction for the rest of the Job Plan.

Chapter 7 describes the Speculation Phase. This phase is the creative process of problem solving. Based on their understanding of the project, the team is challenged to be creative in their thinking. For each function, the team members are encouraged to create numerous ways of performing that function. The chapter provides techniques to increase the number of ideas.

Chapters 8 and **9** describe the Evaluation Phase. This phase sets the stage for the decision process. Two major steps are introduced: screening and judging.

Chapter 8 describes the screening process which eliminates ideas that are not acceptable or needed. The selected ideas will be combined into a meaningful number of alternatives for judging.

Chapter 9 describes the ranking of alternatives by judging. The judging process selects the ideas that warrant further consideration. Judging is based on how the alternatives perform, how they are accepted and whether the alternative is worth the effort. Leading alternatives are then developed for implementation.

Chapter 10 describes the Development Phase, demonstrating how to narrow down the choices so that a preferred alternative can be recommended. It requires technical data and visual aids to sell the ideas.

Chapter 11 describes the Presentation Phase. It summarizes the needs for the project, the options that were explored, reasons for the final selection and how to present them in a simple, attractive and convincing manner.

Chapter 12 discusses two items that are critical to a study: team selection and subject selection. Subject selection uses a specific study to show the selection process. The VE team leader should use this as a guide to develop specific guidelines for a study.

Chapter 13 presents certain features of VE. This includes the concept of VE, timing of VE study, Value Engineering Change Proposal (VECP) and an understanding of team behavior.

Acknowledgements

My professional journey on value engineering began, and continues to revolve, around my work at Alfred Benesch & Company (Benesch). The content of this book is the result of my VE studies, research and development over the past 33 years at Benesch and the contributions of Harold R. Sandberg, Michael N. Goodkind and other Benesch engineers.

My friend, the late Robert K. Carter, talked me into learning value engineering (VE). He thought VE really fit with my philosophy and approach to solving engineering problems. Later, my wife, Sivagami Kasi helped me develop my skills on writing and public speaking so that I could deliver my message. She then joined forces with Harold Sandberg to make sure that I was certified as a Value Specialist. Through my learning process Thomas Cook and Thomas Snodgrass guided and encouraged me to develop my skills on VE.

For the past 15 years Michael Goodkind, has been my technical partner in developing new approaches including Benesch's TQE® program and conducting VE workshops. He continues to complement my skills, and together we are partners in delivering better VE studies to our clients. He also contributed to the technical content of this book.

Any new approaches need to be presented in a good visual format. Jayne Hill and Ken Holt have developed the graphics for Benesch VE presentations for the past 20 years. Jayne also helped me develop the format and content of this book. This book conveys its intent in a precise manner due to the excellent editing by Harold Sandberg and Amanda Rackow.

At the age of 16, I took a course in logic at St. Xavier's College, Palayamkottai, India. Father Chokkiah taught us the power of questioning "Why?". At that age I didn't understand the effectiveness of the logic of questioning. Seventeen years later, learning to ask the question "Why?" gave me the starting boost to understand value engineering. Value will never be achieved if we don't know the reason behind any action.

As you begin to understand the Function Logic Diagram concept of this book, you will learn the importance of basic functions and enhancing functions. All of the people I mentioned above have helped me in the development of my skill and knowledge on enhancing functions of VE.

I credit the person responsible for my basic understanding of the concept of value, my mother, Lakshmi Ammal. She taught me that value in life is beyond the materials we accumulate. It is more of giving, teaching, sharing and enjoying the pleasure that other people gain by our knowledge. Fortunately my wife continues to adopt a similar philosophy that prompted me to write my past publications and this book.

Thank you all.

Muthiah Kasi

Contents

Introduction

Objective

The objective of this chapter is to introduce the VE methodology and VE job plan.

History Of Value Engineering Methodology

Larry Miles developed the concept of value methodology while working at General Electric during World War II. Due to shortage of materials, parts and personnel, products were in great demand. Mr. Miles, when purchasing, approached the problem of shortage by asking the question, "What does it do?" Once he had an answer, he explored the question, "What else will do the job?" Through this process he substituted an equivalent or better product that performed the same function as the one in shortage. This new "function based approach" became very successful and was accepted as a value analysis technique. Larry Miles defined value analysis as "...a system for use when better than normal results are needed."[1]

---

1. Miles, L. D. (1989) <u>Techniques of Value Analysis</u>, 3rd edition. McGraw Hill Book Company, 12.

In the 1960's, Charles Bytheway developed a structure for functions which later became a widely known concept, Function Analysis System Technique (FAST). Later, Thomas J. Snodgrass and Ted Fowler developed the concept of a customer-oriented FAST diagram. The term FAST in this book refers to a Function Logic Diagram.

What is Value?

Value is achieved when the project has a very high performance while reaching a broad acceptance at a reasonable cost. Value is subjective. It reflects people's feeling and their needs at the time. Sometimes people can't define or describe what they want. They confuse needs and desires leading to the perception of a constraint.

People value things differently. Consider this the next time you are in a large parking lot. You will see cars of different types, size, color and price. Each one of those cars was purchased because the owner decided it was the best value for them. As an outsider, we cannot make assumptions about the motivation behind those purchases because we don't know the owners. Don't be surprised if a wealthy person chose an economical car or if a luxury car owner can't quite afford the car payment. Similarly, value engineers should not assume how owners set emphasis on the variables that define value. The VE process described in this book places a high priority on understanding the stakeholders–including the owner, their needs, desires and constraints–in order to determine value.

For example, during a VE study of a train station entrance, the owner insisted on an entrance with the best spatial quality. The team developed the entrance with various shapes and dimensions. The owner kept rejecting each idea as too small, too big, too intimidating or too restrictive. As the team got frustrated, they asked him to define his version of spatial quality. He calmly replied, "I will tell you when I see it." The team proposed more ideas until the owner became excited and said, "That's it!" This example demonstrates how subjective the term value can be and how important it is to involve the owner in this part of the process. When the owners, or their representatives, participate in the study, it helps the team define the value of a project.

Determining Value

Exploring the definition of value is a key task during any VE study. When defining value, value engineers should weigh the circumstances under which value is determined. In addition, they should recognize that the constraints, needs and desires of the users, owners and stakeholders influence value.

Years ago, I arrived at a train station around 5:30 am. The temperature was -20 degrees with a lower wind chill index. I needed 40 cents for a bus ride. I checked my pocket and realized that I had only 39 cents. I checked my wallet for a dollar bill to drop in, but all I had was a $20 bill. I couldn't find a cab anywhere. I ended up walking a mile to the office in the cold weather. I would have used as much as $5 to get a bus ride. The question is "What is my estimated value of a penny?" It is definitely not $20. Under the circumstances, I set an upper limit value of $5. What I learned was that the circumstances often dictate the value, but there is always a limit.

Value by Comparison

Value engineering encourages people to ask the question "Why?" It makes people compare the benefits of various options and determine their worth. This approach can be practiced within the context of every decision in life. When my son, who was living in Illinois, wanted to go to an out-of-state university, I asked the question, "Why do you prefer to go to an out-of-state university?" His answer indicated that he wanted to go school with some of his friends. Out-of-state tuition would have cost $9,000/year ($36,000 for four years). For my son the out-or-state school would accomplish the function of *Feel Comfortable*. Instead of telling him it would cost us an additional $36,000 we equated that to a cost of a sports car. We promised him a car upon graduation if he chose the in-state university. He was given a choice of two functions: *Feel Comfortable* vs. *Receive Gift*. In the end, he chose the Attract User function over Assure Convenience function by opting for the in-state school. When he graduated from school, his accomplishment and his maturity made him decide he didn't want the sports car from us. He understood later that his decision to stay in-state resulted in a much higher value. My wife and I were not surprised with the outcome. Similarly in projects, value engineers should look for ideas with alternative functions to compare. Decision makers will be able to understand appropriate value comparisons and will be able to better appreciate its long term use.

Consider an interchange reconstruction project with two alternatives: one would keep all ramps open for a three year construction period, the second would require the ramp traffic to be rerouted with a two year construction period. Which alternative has more value, more inconvenient for a shorter period or less inconvenient for a longer period? These questions are very subjective. Under various circumstances both alternatives are acceptable. Defining value is a critical part of the VE process and instrumental to the ultimate satisfaction of the stakeholders. One way to define value is by comparing various alternatives.

What is Value Engineering?

Value engineering is a systematic team approach to creatively enhance the value of a project or product. In the industry, VE is often referred to with various other names such as value analysis, value planning and value management. When the technique is applied during project planning, it is called value planning. If the techniques are used on an already designed or constructed project, it is called value analysis. The approach to each of these varies slightly, however the job plan remains the same.

VE Job Plan

Value engineering advocates a step-by-step, systematic approach to solve a problem. This systematic approach is embodied within a job plan that has several phases and imposes a set of rules that must be adhered to. The rules may appear to be simple, but they are vital to the success of the value engineering study. This section describes a typical job plan and explains the rules of the job plan and the reasoning behind them.

A typical job plan consists of the following phases:

Job Plan	
	▶ Information Phase
	- Function Analysis Phase
	▶ Speculation Phase
	▶ Evaluation Phase
	▶ Development Phase
	▶ Presentation Phase

These phases are illustrated below and are described in more detail in the chapters that follow. These phases can also be summarized with a series of questions, also presented here.

Information Phase

The Information Phase is divided into two sections. The purpose of the first section is to understand the user, owner and stakeholders, and their constraints, needs and desires. The second section is to translate this information into functions of the project. This includes a unique VE technique, function analysis. Function analysis lists the functions, places them in a logical manner and allocates cost to each of them. High cost functions will be tested for their worth. This will be the basis for the speculation of ideas.

Critical questions answered:
What is it?
What does it do? (functions)
What does it cost?
What should it cost? (function cost)

To define functions of the project, the VE team needs to know the owner's/designer's intended solution. The status of this solution depends upon the stage of the project. The following are suggested terms to use to convey the status of the project:

Phase	Designation
During Planning	As Given
During Design	As Intended
At the Completion of Design	As Designed

The VE alternatives generated by the VE team should be designated As Proposed.

Speculation Phase

In this phase, the VE team explores alternative ways of performing functions that enhance, or at least maintain, performance or acceptance at a reasonable cost.

The key to speculation is to let the mind wander freely, with no limitations. Several techniques can be used to assist the participants in exploring the widest possible range of ideas. These are described in Chapter 7.

Critical questions answered:
What else will do the job?
What if... ?
What is the least expensive way of performing each function?

Evaluation Phase

In the Evaluation Phase the VE Team will rate and rank feasible ideas. The evaluation phase is performed in two steps, screening and ranking.

Critical questions answered:
Will it work?
Will it be acceptable?
Can we afford it?

Development Phase

In this phase the VE team will add information that will facilitate selection of a preferred alternative among the highest ranked alternatives in the Evaluation Phase.

Critical questions answered:
How can we make it acceptable to owners/users?
What can we do to make it better?
What can be done to make it cost less?
What are the strengths and weaknesses of the leading alternatives?

Presentation Phase

In the Presentation Phase the VE team prepares to convince decision makers to accept the study results. This is accomplished through an effective presentation with facts and features.

Critical questions answered:
How do we present our recommendations?
What are the road blocks?
Who really makes the decision?

Rules for the VE Job Plan

The rules that govern each phase of a value engineering study are briefly described next.

Information Phase

- Do not speculate
- Do not jump to conclusions. Gather as much information as possible
- Do not evaluate
- Understand the problem

Speculation Phase

- Do not bring preconceived notions to the discussion
- Do not question the validity of an idea
- Encourage everyone to participate in the creative phase

Evaluation Phase

- Do not create
- Do not jump to conclusions
- Judge each idea against the criteria
- Always be prepared to explain your ratings

Development Phase

- Improve ideas
- Double check that assumptions are reliable

Presentation Phase

- Do not assume that ideas are good
- Demonstrate their worth
- Follow these three steps: inform, instruct and influence

Conclusion

In this chapter, an overview of the value engineering methodology was presented. The questions listed under each phase provide foundations for the understanding of the phase. Rules were presented for each phase. These can be used as checkpoints to ensure the overall objectives for that phase are being met, while adhering to the appropriate parameters. The next chapter describes the first part of the Information Phase: Understanding the project and the stakeholders.

Information Phase

Objective

The objective of this chapter is to understand the purpose and need of the project. Needs and desires of the stakeholders, owners and users are also discovered. The team needs to understand the problem, the given solution, the stakeholders affected by the problem and the expectations of these stakeholders. This understanding will later be translated into project functions. Project functions are described in the Chapter 3.

Introduction

"If I had only known what I know now, I could have saved a lot of time and pain."

We have all heard similar statements many times in our lives. Exploration of ideas or events prior to the conclusion is key to a better decision process. This is exactly what value engineering process expects in its job plan. Understanding the existing conditions, exploring people's needs and desires are all a part of the information gathering phase. The job plan for VE begins with the Information Phase. In this phase the following questions are asked and answered:

What is it?
What does it do?
What does it cost?
What should it cost?

These questions can be addressed by describing existing conditions, understanding the issues or the problems, identifying the people behind the issues, discussing the As Given solution, if any, and estimating the cost of the solution. In addition, the question "What does it do?" is addressed in terms of functions. The concept of function is the backbone of the VE process. This process is explained in detail in Chapter 3.

Process

In order to ensure that all of the appropriate information is gathered, the following steps are listed as a guide. Explore each of these steps in detail and document the results. This list is only a guide, feel free to explore other areas that may uncover information useful to the project.

Information

Step 1. Describe project including the project history (Chapter 2)

Step 2. Identify stakeholders and develop constraints/needs/desires (Chapter 2)

Step 3. Determine the project functions from the list of stakeholder constraints, needs and desires (Chapter 3)

Function Analysis

Step 4. Develop a Function Logic Diagram from the perspective of the project. (Chapter 4)

Step 5. Determine the cost of each element. (Chapter 5)

Step 6. Distribute the cost of each element to each of its functions and analyze its worth. (Chapter 6)

Stakeholders

A key component of the Information Phase is to identify the stakeholders. Stakeholders are the ones who determine if the project is a success or failure.

There are three kinds of stakeholders: users, owners and others.

Owner	▶ Financially responsible for funding the project ▶ Shares in the funding ▶ Represents the owner's interests ▶ Manages the project

User	▶ Actively uses or maintains the project

Others	▶ Financially affected by the project ▶ Environmentally concerned about the project ▶ Disturbed by any required change in habits or recreation

Define Stakeholder Constraints/Needs/Desires

Each stakeholder has expectations regarding the project. These expectations are grouped into constraints, needs and desires. Descriptions of these are as follows:

Constraints	▶ Legal requirements ▶ Standards of the owner ▶ Physical conditions of the site ▶ Commitments to stakeholders

Needs	▶ Expectations that must be fulfilled by the project if constraints are not violated ▶ Limitations or restrictions that are imposed by stakeholders but which can be violated. The degree of violations will be considered in the evaluation of alternatives.

Desires	▶ Expectations that should be fulfilled if cost is not a factor.

There are several points to keep in mind when identifying the stakeholder constraints, needs and desires. First of all, the majority of constraints are dictated by the law and the applicable codes and standards for land use. These constraints are too numerous to be listed for each VE Study.

The constraints, needs and desires that should be listed are those imposed by a stakeholder or by a special code or standard. For example, in upgrading an interstate highway, the *"AASHTO Policy on Geometric Design of Highways and Streets"* calls for a minimum vertical clearance over the interstate of 16'-0". However, if the improvement is limited in scope, it is possible to reduce the required minimum vertical clearance to 14'-6" or to even maintain the existing vertical clearance, if it is less than 14'-6".

Design criteria should be described as a constraint, need or desire. For example, some stakeholders may say that vertical clearance over an expressway might be shown as a constraint of 14'-6" (to meet most urban requirements), a need of 16'-0" (to meet AASHTO) and a desire of 16'-3" (to account for future overlays). On the other hand, other stakeholders may say that 16'-3" is the constraint (no design exceptions), the need is 16'-0" (to meet AASHTO) and the desire is 14'-6" (to reduce the cost of the improvement).

Needs and desires are generally not executable. They are more like visions of what the project should do. Sometimes, ideas are mistakenly offered in lieu of needs or desires. Ideas are executable. The reason behind any idea should be listed as a constraint, need or desire. However, the stakeholders usually do not follow this guideline. In the case study, the stakeholder statements are listed as expressed in a public meeting. The VE team converted the reasons behind their statements into project functions.

Case Study – STH 34 Bypass Information Phase

As described in the preface of this book, one project case study will be used to demonstrate each of the VE phases. Application of the ideas presented in each chapter allow for a clearer understanding of this methodology. The case study used was an actual Benesch project—the identifying names and labels have been changed. Let's begin by exploring the Information Phase Process.

Description of Project

STH 34 in Wisconsin was originally intended to be a bypass around the City of Coldwater. However, local developments over the last two decades made the bypass into a local roadway. This resulted in conflicts between the thru traffic and the local traffic. Exhibit 2.1 shows the existing roadway. In addition, the planned growth along nearby STH 34 was substantial. The potential growth included the possibility of a new high school on the east side of STH 34

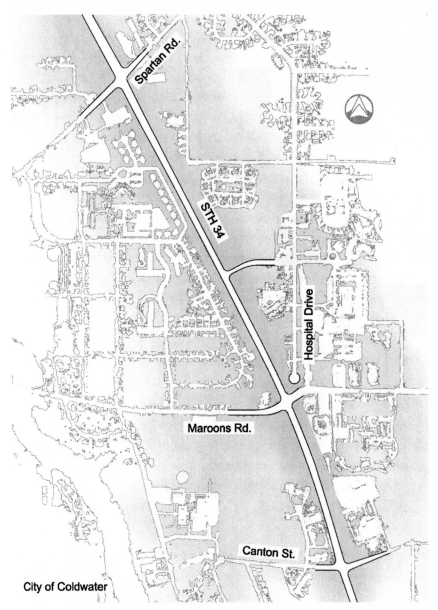

Exhibit 2.1: Location map STH 34

along Spartan Road; a shopping mall west of STH 34 along Canton Street; and an facility expansion to the hospital located on the east side of STH 34.

Public concern had been expressed for the safety of local residents crossing STH 34 against high-speed thru traffic. Public concern reached a peak after the tragic death of a young girl who was crossing STH 34 at the Maroons Road crossing.

Also of concern was the number of reported accidents along STH 34. The current rate was comparable to those on an urban highway. This level was perceived to be too high for this rural roadway. The objective of the Wisconsin STH 34 Bypass study was to develop a plan and strategy for safe handling of the current and future local and thru traffic.

Identify Stakeholders

Given the nature of the issues, this project was a stakeholder driven study. The bypass was built by the state to serve the thru traffic, but since that time the city expanded around the bypass and the local population used the bypass as a local road. In order to identify the stakeholders, a list was created that considered every person that might have an opinion on this project. Understanding the role of the following stakeholders was the first step in solving the issues of the community:

1. Wisconsin Department of Transportation
2. City of Coldwater
   a. City Council
   b. Engineering Department
   c. Fire Department
   d. Police
   e. Residents
3. Real estate business

4. Adjacent property owners
5. Local traffic
   a. Hospital
   b. Apartments
   c. Future school
   d. Proposed mall
   e. Local business
   f. Pedestrians
   g. Bicyclists
6. Thru traffic
7. McArthur County

Define Stakeholders' Constraints/Needs/Desires

Constraints, needs and desires are in the point of view of the stakeholders. They sometimes contradict each other. The following list details the key constraints, needs and desires (C/N/D):

We want the road to be very safe.	C
Don't reduce the speed. This is a bypass.	N
Reduce the speed. This is a local road.	D
Put a bridge to go to the hospital. We don't want to go around.	N
We should be able to cross wherever we want.	D
We should restrict pedestrian crossing.	D
We should have better access across the road.	N
It is taking too long to cross the bypass.	D
They are planning a school. The traffic will be worse.	D
We don't want to stop suddenly when we travel at 60 mph.	N
This is a local road. We should be able to use it as we want.	D
Don't put too many bridges, it will look like a city.	D

We don't want our property taxes to go up because of the construction.   D

We don't have funding allocated for the project.   C

The state has about $4 million for this project.   C

Don't take any of our property.   C

We need new developments. This highway should accommodate them.   D

Don't tie-up our roads all summer with construction.   D

We have too many accidents. Do something.   C

We want the city engineer to solve this mess.   N

These are stakeholder constraints, needs and desires which can cause conflict with each other. In this phase it is important to understand the stakeholders.

For years there was no proposed solution that satisfied all the stakeholders. Although all stakeholders were interested in an acceptable solution, a solution never surfaced. Exhibit 2.2 summarizes the conflicting needs of major stakeholders.

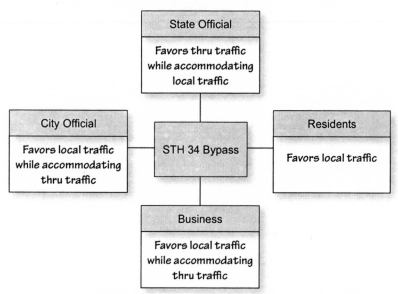

Exhibit 2.2: Chart summarizing conflicting needs of stakeholders

This study took tentative steps in understanding the conflicting needs, respecting the stakeholders constraints and searching for an acceptable solution. This information was the basis in the development of the functions of the project within the framework of the stakeholders' needs, desires and constraints.

Conclusion

What we learned in this chapter is that, at times, the designers are faced with strong and influential stakeholders with justified needs that conflict. A thorough understanding of these needs, a true respect for their concerns and the offering of creative solutions are critical to a good study. The case study used in this book is a classic example that demonstrates the effectiveness of value engineering. Local city residents need a local road for their daily use. The thru traffic needs a freeway that will help them pass through with minimum delays. Value of the improvement will be achieved with a balanced solution that addresses the needs of all the stakeholders.

Still within the confines of the Information Phase, the next chapter describes how the stakeholders' needs and desires are matched with the project needs, which are functions of the project.

Information Phase–Functions

Objective

The objective of this chapter is to express the needs and desires of the stakeholders in terms of functions that the project should perform.

Introduction

Functions are defined in two words: an active verb and a descriptive noun. For example, *Increase Safety*. It is also important that the functions are expressed from the project point of view. The *SAVE International Monograph of Function* states "The determination of function(s) is a requisite for all value studies"[1] and "All cost is for function."[2]

It also states that "A function is that which makes an item or service work or sell—in other words, an item's function is why

---

1. Miles, L. D. (1989) Techniques of Value Analysis, 3rd edition. McGraw Hill Book Company, 12.

2. Macedo, Dobrow and O'Rourke. Value Management for Construction, John Wiley and Sons. 242.

the customer buys the product or service."[3] An item, or in the case of transportation, structures or highways, is a means to an end of providing a function, not the end itself. In using the function approach, the value study team must remain focused on the primary reason for design and build cycles, which is the ultimate use of the item. Stakeholders accept a service or a design because it will provide a function that satisfies their need at a cost they are willing to incur.

Functions

The most important feature of value engineering is function analysis. The concept of functions must be clearly understood in order for one to get the most out of value engineering. It is not enough to capture the general idea of the project, but to exert the necessary effort to correctly identify all the functions.

Larry Miles created the value analysis concept by asking the simple question: "What does it do?" The answer to this question defines the functions of the project.

Mr. Miles stressed that functions, not parts of a project, should influence the creation of alternatives. The ability to satisfy these functions leads to increased value. The importance of first determining the right functions is illustrated by the following example.

Consider a branded coffee mug that displays messages on it. What does this mug do?

- Is it a receptacle for liquid intended for consumption?
- Is it a type of media on which greetings are displayed?
- Is it a paper weight?
- Is it used to hold smaller items, such as candy or coins?

The functions are greatly different for each of these questions. Responses to these questions would be *Facilitate Consumption; Convey Message; Secure Paper; and Collect Items.*

3. SAVE International. (1998) <u>Monograph Function: Definition and Analysis</u>, 3.

Each of these are possible functions for the mug. For value engineering, we need to identify which one of these function is most closely related to the business decision at hand. Understanding and identifying this would lead to the ultimate selection of the primary function, and set the direction for the study.

What is a function?

- A function is a required action described by a verb (active) and a noun (descriptive) without identifying a specific method of performing that action. One or two adjectives can be used to modify the noun.

- Functions enable us to focus on what the project or the element does, rather than on its components.

- The needs and desires of the stakeholders as seen from the project point of view.

Process

The most effective approach for function identification is to use the component/function form.

- List the parts, process or components
- Subdivide the parts, process or components into smaller parts
- Assign functions to these subdivided finite elements or parts

For the coffee mug example, the mug is divided into the following parts with the appropriate functions.

Part	Function
Mug	*Hold Liquid*
Handle	*Facilitate Handling, Protect Holder*

Material	*Insulate Heat, Protect User*
Information	*Convey Message, Advertise Company*
Color	*Attract User*

Guidelines

When defining functions, follow these simple steps:

- Use a dictionary or thesaurus to get precise definitions of function words
- Avoid using the name of the part or action as a function
- Avoid using generic verbs such as "provide" or "furnish"
- Use active verbs and descriptive nouns

Example - Locks in a Door

When analyzing a door as a part of a VE study, *Provide Lock* would be a weak function to list. This description limits the creative options and excludes other types of functionality. If the function is defined as *Secure Home* or *Secure Office* it opens up various options including but not limited to the following:

1. Install dead bolts
2. Install security bars
3. Install electronic keys
4. Set up security systems
5. Install cameras
6. Hire security guards
7. Inform police when you go away
8. Be friendly with your neighbors
9. Do not inform your neighbors
10. A combination of the above ideas

Provide Lock is limited to the first three options, whereas the function *Secure Home* applies to all 10 of the options. The list can be expanded. The message is to define your function such that it opens up various options. Functions can limit or expand the possibilities.

Case Study – STH 34 Bypass Functions

For the STH 34 case study, let us explore the functions. Both the state and city transportation officials agreed on certain functions that related to capacity, access and safety of local and thru traffic. However, they disagreed on how to accomplish it without affecting traffic. The following is a list of functions related to capacity, access and safety:

1. Manage Traffic
2. Safeguard Thru Traffic
3. Safeguard Local Traffic
4. Maintain Local Access
5. Reduce Delays
6. Maintain Speed
7. Safeguard Pedestrians
8. Safeguard Residents
9. Control Thru Traffic
10. Control Local Traffic
11. Control Pedestrian Traffic
12. Separate Traffic
13. Guide Traffic
14. Minimize Indirection
15. Channelize Traffic
16. Maintain Access

17. Maintain Local Traffic

No matter what the solution, it needed to be aligned with long term plans for both the city and state. This resulted in additional functions:

18. Satisfy Local Long Term Plan
19. Satisfy Regional Long Term Plan

The residents, businesses and political officials realized there would be substantial construction that could interrupt daily operations. From this acknowledgement came additional functions:

20. Minimize Right-of-Way (ROW) Impacts
21. Minimize Construction Time
22. Preserve Local Road Concept
23. Encourage Community Cohesiveness

Local concerns and public perception related to growth and safety needed to be addressed. The following are functions that relate to this perception:

24. Address Safety Perception
25. Attract New Development

Each of the functions listed here reflect the stakeholders' needs and desires. The ultimate project improvement should satisfy these functions. Through the development of a Function Logic Diagram, missing functions will be realized. It is normal to add functions at that stage. In the next chapter we will discuss this process.

Once the functions are listed, the next step is to place them in a logical order. What is the basic function or the task? How does one distinguish between a secondary function and a primary function? Several methods have been used, but the real break through came in 1965 when Charles Bytheway presented his paper, "Basic Function Determination Technique," at the Fifth Annual Society of American Value Engineers Conference in April 1965. This is the subject of the next chapter.

Conclusion

Defining functions and understanding them differentiates value engineering from other cost-reduction techniques. Embracing the technique of function definition is the most difficult adaptation an engineer must make to become a value engineer. It is hard to force oneself to do it. But when properly done, it enables one to understand the scope and objectives of a project.

Information Phase—Function Logic

Objective

The objective of this chapter is to classify functions and group them into a logical format for better understanding.

Introduction

Function analysis is the second part of the Information Phase. Functions convey the purpose of a project. When these functions are displayed with a How-Why logic, the reason for the project becomes clear. The logic diagram focuses on a single reason for the entire project. The function that is the driving force is termed the *task* or *higher order basic function*.

The first part deals with constraints, needs and desires of the stakeholders. In management terms, it is the result of lateral thinking. In lateral thinking, people tend to address their own issues without worrying about another people's needs or desires. It does not mean that they don't care. Sometimes one person's needs conflict with other person's needs so they are not easily recognized. In function analysis, the VE team will bring these needs and desires into the project point of view.

In the last chapter we learned about the formulation of functions of the project and how they describe the purpose and needs of the project. The next step is to sort these functions by priority or importance. Keep in mind that most of the functions listed by the stakeholders are important to the stakeholders. However, the team must know the role each function plays in the project. First, the team needs to select the one function that is the driving force of the project. In a highway improvement project, the team must know the main function for the proposed improvement. Is it about capacity, safety, level of service, future development, stakeholder convenience or a combination? In this chapter, this main function is classified as a *task or higher order basic function*. Once the task is identified, the next step is to identify the very basic functions that satisfy the task. The rest of the functions are classified as enhancing or supporting functions that complete the task. For example, if *Improve Safety* is a basic function, *Maintain Speed* is an important enhancing function. By identifying the purpose of each function, various alternatives can be evaluated with proper weight of importance.

Function Analysis System Technique (FAST)

FAST is a diagramming technique that specifically illustrates the relationships and interrelationships of all functions within a specific project, using a "How–Why" logic pattern. Initially called Customer FAST, the diagram is also known as Customer-Oriented FAST Diagram. Recently, the name FAST diagram was replaced by the term Function Logic Diagram. In this book we will refer to it as Function Logic Diagram. There are two Function Logic diagram variations.

This chapter introduces the two types of function logic diagrams to classify functions: the Technical Function Logic Diagram and the Customer-Oriented Function Logic Diagram. In most cases they can be used interchangeably. One type favors customer or stakeholders needs whereas the other favors the designers point of view, with emphasis on stakeholder needs and desires.

Technical Function Logic Diagram

Technical Function Logic diagramming is effective when dealing with an element within a project. The situation or element is an assembly or a portion of a critical function path. Terms or functions are oriented to technical activities. A Technical Function Logic Diagram has a specific structural form (Exhibit 4.1).

There are four important elements in a Technical Function Logic Diagram:

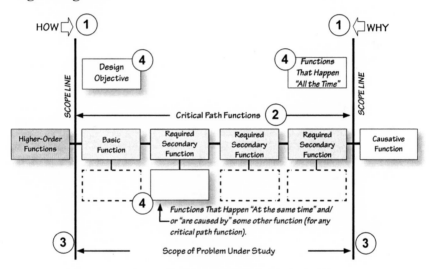

Exhibit 4.1: Technical Function Logic Diagram

Element 1: "How-Why" Logic Questions

Function analysis requires analyzing why a function exists and how a function satisfies other functions to complete the link between them. This "How–Why" logic assures that all the required functions are listed in the Function Logic diagram.

Element 2: Critical Path Functions

The critical path begins with the higher order function which defines the need of the project. It ends with the causative function. It is called a causative function since it starts the critical path to the basic and higher order function. Determining the basic function often requires selecting functions from the list of suggested functions and applying the "How" and "Why" questions. If the "Why" question is answered by another identified function, that function is the next candidate for the basic function. The function to the right becomes a required secondary function. Once the basic function is verified, the remaining required secondary functions are identified. This group makes up the critical path.

In the example of the cup, the purpose or the main driving force may be different for different users. It can be *Hold Liquid* or *Convey Message*.

Task or the higher order function *Convey Message* may lead to the ideas shown in Exhibit 4.2.

Exhibit 4.2: Ideas to Convey Message

Task or the higher order function *Hold Liquid* may lead to the ideas shown in Exhibit 4.3.

Exhibit 4.3: Ideas to Hold Liquid

It is important to understand the purpose of the project. In a highway project the higher order function may be capacity driven, safety driven or future development driven. By selecting the appropriate higher order function one can achieve a solution that will have a good value.

Element 3: Scope Line

To study in detail the functions affecting the selected elements of the project, place a vertical scope line to the right of the highest-order function performed by the selected element. The function to the right of this scope line is the basic function of the selected element. Place a scope line to the left of the lowest critical function performed by the relocated element. This is the causative function since it really starts the critical functions. Determine which of the required secondary functions is the basic function by using the "How–Why" logic on all the functions between the scope lines.

Element 4: Non-Critical Path Functions

The last group of functions are enhancing functions. There are three types. The first type, "caused by" or "same time" functions, connect directly to a critical function. These functions result from the performance characteristics of particular critical functions and act as modifiers. The second type, "all-the-time functions," modifies two or more of the critical functions. The third type, "design objectives," represents specifications that are added to the design, often by the stakeholder or group that is developing or operating the process.

Example - Establishing Scope Line

The scope line will set the direction of any VE study. To set the scope line, collect the functions related to the study. Let us take an example of function approach to determine a good method of presentation. The functions related to presentation are

- *Influence Decision Makers*
- *Convey Message*
- *Illustrate Data*
- *Develop Slides*

Pick any one of the functions and ask the question Why? and How?

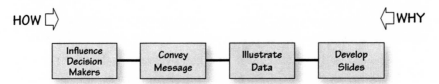

If the scope line is left of the function *Develop Slides* then the higher order function is *Illustrate Data*.

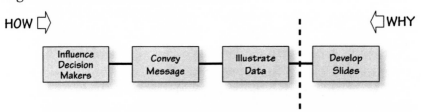

In this case, the design objective is to *Impress Audience. Influencing Decision Makers* is beyond the scope.

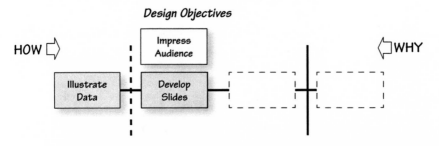

If the scope line is left of *Convey Message* then the higher order function is *Influence Decision Makers*.

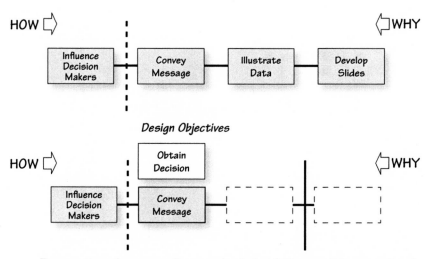

By moving the scope line to the left of *Convey Message*, the design objective is broadened to the design objective *Obtain Decision*. In this case, *Illustrate Data* is one of the critical functions in the critical path scope line placement is very sensitive to define the scope.

Case Study - STH 34 Bypass Technical Function Logic Diagram

Functions of the STH 34 Bypass are presented in the form of technical Function Logic in Exhibit 4.4. Functions that are listed in the previous chapter were modified during the placement of the functions in the logic diagram. Sometimes functions are combined, deleted, divided or modified at this stage. This is the result of the "How–Why" logic exercise. For example you will notice *Safeguard Pedestrians* and *Safeguard Residents* are combined into one function, *Safeguard Residents* by eliminating the other function.

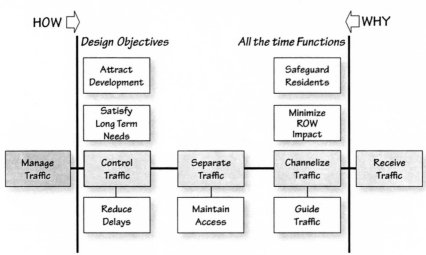

HOW ⇨   ⇦ WHY

Design Objectives   *All the time Functions*

| Attract Development | Safeguard Residents |

| Satisfy Long Term Needs | Minimize ROW Impact |

| Manage Traffic | Control Traffic | Separate Traffic | Channelize Traffic | Receive Traffic |

| Reduce Delays | Maintain Access | Guide Traffic |

Exhibit 4.4: Technical Function Logic Diagram of STH 34 Bypass

Higher Order Function

Try to assemble the critical path functions first. From the list of the functions pick one as the most critical function. Either *Manage Traffic* or *Control Traffic* seems to fit with the need of the project improvements. Let us try *Manage Traffic*. Then apply the How-Why logic.

How do you *Manage Traffic?* The answer is *Control Traffic*. At present, thru traffic and local traffic use the Bypass without any imposed restrictions. This resulted in conflicts and accidents.

How do you *Control Traffic?* The answer is *Separate Traffic*.

Traffic can be separated by various means including signals, stop signs, grade separation or by police. The function will be the basis to speculate at the next phase.

How do you *Separate Traffic?* The answer is *Channelize Traffic*. How do you *Channelize Traffic?* The answer is to *Receive Traffic*.

The critical path logically described the steps of *Managing Traffic*. Traffic in all directions is received, channelized, separated and controlled.

The critical path is formed with a series of How questions from left to right. The logic should be tested by asking a Why question from right to left.

Why do you...	The answer is to...
Receive Traffic?	*Channelize Traffic*
Channelize Traffic?	*Separate Traffic*
Separate Traffic?	*Control Traffic*
Control Traffic?	**Manage Traffic**

Design Objectives: *Satisfy Long-Term Needs*

In the past, the proposed alternatives were rejected because they conflicted with the regional planning. In this study, function reflecting this issue becomes the design objective.

All-the-time functions: *Safeguard Resident* and *Minimize ROW Impact*

Alternatives are expected to fully address the safety issue. In addition, the city and its residents are very much concerned about the impact on their properties. These two issues should be addressed for all alternatives.

Secondary Functions

Now let us explore the secondary functions, caused by functions and functions that happen at the same time. Whenever traffic is controlled, it can cause delays. Critical path function, *Control Traffic* caused the creation of a "caused by" function *Reduce Delays*. Similarly, when traffic is separated, access may be limited or eliminated. At present, local traffic enters the Bypass with no restrictions. Separation of the traffic may increase the local traffic route. This resulted in the creation of the function *Maintain Access*. As the traffic is channelized, it should be properly guided: *Guide Traffic*. This is an example of "Happens at-the-same-time function."

All alternatives that will improve the traffic management will satisfy the critical path functions. Each alternative satisfies other functions with a different degree and at a different cost. Prior to evaluation, alternatives will be examined for its fulfillment of these functions.

Customer-Oriented Function Logic Diagram

The Customer-Oriented Function Logic Diagram logically displays stakeholders needs and desires (Exhibit 4.5). Customer-Oriented Function Logic diagramming is especially effective in the planning or conceptual phase. Use conceptual layout and building plans to develop these Function Logic Diagrams.

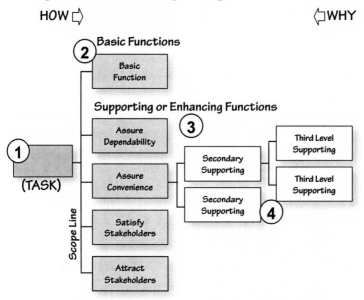

Exhibit 4.5: Customer-Oriented Function Logic Diagram

There are four elements to the Customer-Oriented Function Logic Diagram:

Element 1: Task

The first step is to determine the task. The task satisfies the overall purpose of the project. Establish a scope line just to the right of the task. Functions that answer "Why perform the task?" lie outside of the scope. The concept of establishing the scope line is similar to the one discussed under Technical Function Logic Diagram.

Element 2: Basic Functions

The second step is to separate the identified functions into basic and supporting functions. Basic functions are those which are essential to the performance of the task. Without the primary basic functions, the project or process will not work.

Element 3: Supporting or Enhancing Functions

The third step is to group the remaining functions into the four primary supporting function groups. Supporting functions play an important role in a building. Structural engineers, for instance, concentrate primarily on the basic functions, with heavy emphasis on the primary supporting function *Assure Dependability*. Mechanical engineers and electrical engineers pay more attention to the supporting function *Assure Convenience*, while architects satisfy the basic and supporting functions *Satisfy Stakeholders* and *Attract Stakeholders*.

Assure Dependability—Any function that assures dependability has at least one of the following attributes:

- Makes the elements of the project stronger or more reliable or effective

- Makes it safer to use

- Lengthens the life of the parts or minimizes maintenance cost, or both

- Protects the environment

Assure Convenience—Any function that assures convenience has at least one of the following attributes:

- Modifies the basic function to make it convenient to use

- Enhances spatial arrangements

- Facilitates maintenance and repairs

- Furnishes instructions and directions to stakeholders

Satisfy Stakeholders —Any function that satisfies stakeholders has at least one of the following attributes:

- Modifies the basic function to satisfy individual desires

- Makes the stakeholder's life more pleasant. *Minimize Noise* would be an example

- Makes the element appear to be better in the opinion of the stakeholder, but not necessarily in the opinion of the designer. (Sometimes these opinions are reflected in the standards and specifications of a particular agency/owner.)

Attract Stakeholders—Any function that attracts stakeholders has at least one of the following attributes:

- Emphasizes the visual aspect or other senses

- Projects a favorable image (that is, trademarks or endorsement by public figures)

Element 4: Classify Functions

The fourth step is to classify the functions as primary, secondary or tertiary (third level).

The link between the task and basic functions is the sequence of the logical question "How–Why." The "How–Why" concepts must work between the selected task and the primary basic functions. These primary basic functions are interdependent and both are essential to the performance of the task.

Once the primary basic functions have been identified, the question "How" can be asked of each of them. Functions that answer the question "How" will be found in the expanding branches. These are the secondary basic functions. There must be two or more secondary basic functions to justify branching from the primary function.

In a similar manner, the secondary supporting functions branch to the right from the primary supporting functions when the question "How" is applied. Again, there must be two or more secondary functions to justify branching.

This rule also affects further branching off to the third (tertiary) level. Usually, the tertiary level completes the branching basic functions. The end of the branching is obtained when the hardware description or action is the noun of the function. The branches must also satisfy the "Why" question in the opposite direction, that is, the logic check.

Case Study - STH 34 Bypass Customer-Oriented Function Logic Diagram

For the STH 34 Bypass project the functions were listed. The driving force was assumed to be the function, *Assure Safety,* because of the fatal accident. However, it is the emotional response to a tragedy. Everyone agrees that safety is the most important issue. However, the state may respond by prohibiting pedestrians in the proximity of the highway. That will definitely assure safety. But the city may take the position that safety can be assured by slowing the thru traffic. The real issue is how to manage the local and thru traffic to assure safety. This leads to the conclusion that the function *Manage Traffic* is the task and safety-related function, listed under Assure Dependability, is an enhancing function. The readers will appreciate this logic when they review the final conclusion in the later chapter.

Functions listed in the previous chapter were also modified during the placement of the functions in the logic diagram. Sometimes functions are combined, deleted, divided or modified. This is the result of the "How–Why" logic exercise. For example you will notice *Safeguard Residents* was branched into two functions *Safeguard Pedestrians* and *Safeguard Local Traffic*. The main objective was to convey the purpose and needs of the project and classify them appropriately.

All functions related to safety are listed under dependability functions, all functions related to speed or access are listed under convenience, all functions related to planning and construction are listed under satisfy stakeholders and attraction functions are listed under attract stakeholders (Exhibit 4.6).

Development of Concepts

The Function Logic Diagram considers every stakeholder's interest and defines the task and basic functions based on what is the need of the project. However, it does include the stakeholder's needs and desires in the four enhancing function categories. Specifically they are listed under satisfy stakeholders if they don't have a logical reason to fit under the other three categories. To facilitate the next phase, the Speculation Phase, concepts based on the function logic exercise will be listed.

In the Speculation Phase, random ideas will be generated. Later they will be grouped under concepts to assure that ideas favoring all concepts are generated.

HOW ⇨

⇦WHY

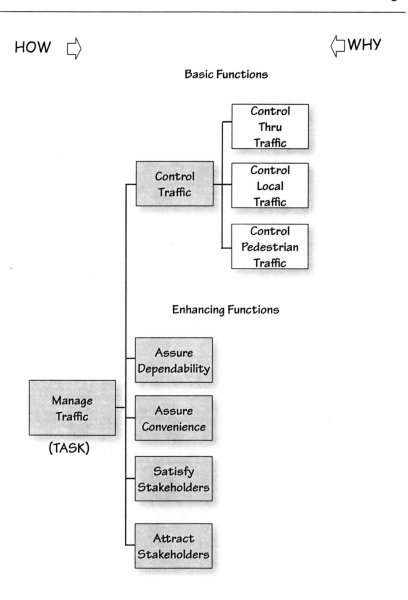

Exhibit 4.6: Customer Function Logic Diagram for STH 34 Bypass

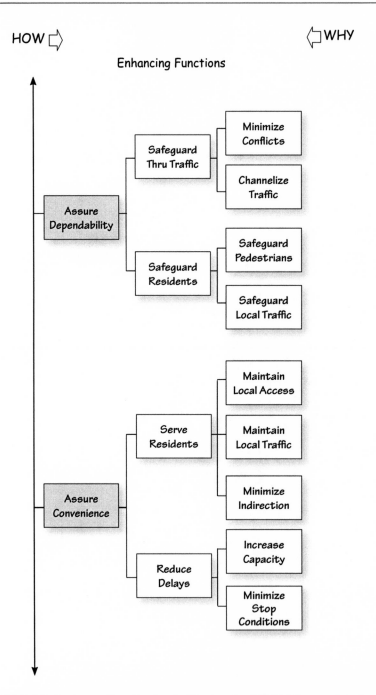

HOW ⇨   ⇦WHY

Enhancing Functions

Exhibit 4.6: Customer Function Logic Diagram for STH 34 Bypass (continued)

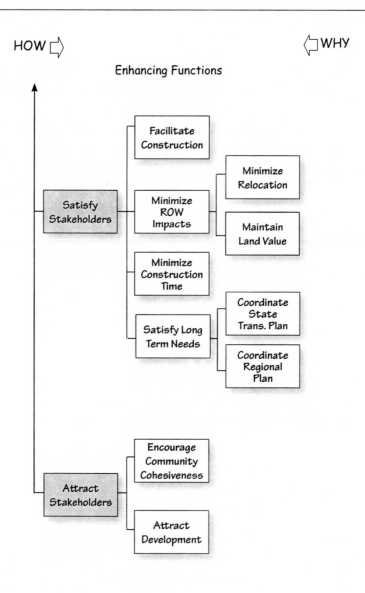

Exhibit 4.6: Customer Function Logic Diagram for STH 34 Bypass (continued)

Conclusion

In this chapter we learned the most important technique of VE. By arranging all the functions in a logical manner, the focus of the study is achieved. In both diagrams, the focus of the effort is to *Manage Traffic*. The study usually provides a solution to analyze. The Function Logic Diagram reflects this solution. At the planning phase the Customer-Oriented Function Logic Diagram is applicable. During the Design Phase design details may fit better with the Technical Function Logic Diagram. Keep in mind that either diagram can be used at any stage of the project.

The next chapter discusses the estimation of project cost. The project cost will then be used in cost allocation to various functions.

Information Phase–Cost

Objective

The objective of this chapter is to introduce the use of elemental cost as opposed to material cost or trade cost. Elemental cost helps to allocate cost of parts to functions.

Introduction

The cost of the project is used to determine the value of the given alternative of the project in the Information Phase. Later it will be used to determine the value of the preferred alternative. Cost is also used as a guide to determine the level of effort required when analyzing specific elements. In the planning stage it is difficult to quantify cost since the elements of the project are not clearly defined. Experienced design team members will help the VE team assume possible elements and help develop a cost estimate. There are four ways to develop estimates: 1) Order of magnitude, 2) Square foot and linear foot, 3) Systems or assemblies, 4) Unit price.

Order of Magnitude

In the program level of a project, the order of magnitude is used. For example, based on the magnitude of the project, the cost of an interchange can vary from $25 million to $60 million. The cost of grade separation structures varies from $1 million to $3 million. Later when bridges and roadway are defined, the square foot and linear foot method can be applied.

Square Foot and Linear Foot

When the length of roadways, number and geometry of the bridges are defined, the estimate can be expanded to be more specific. Square foot cost of grade separation varies from state to state. The Department of Transportation has established values for different type of structures and roadways. For a bridge of 45' wide and 99' long cost can be calculated by established square foot cost of $160 per square foot.

For a bridge carrying a two lane roadway over a two lane roadway, estimate the dimensions of the bridge with a square foot of the bridge (Exhibit 5.1).

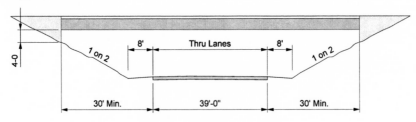

Exhibit 5.1: Bridge elevation

Length = 2 * 12' lane + 15' median + 2 *30' clear zone = 99'
Width = 2 * 12' lane + 2 *8' shoulder + 2' + 2 * 1.5' barrier = 45'
Estimated cost = 45 x 99 x $160/square feet = $712,800

Systems or Assemblies

During the schematic level of a project, cost can be estimated in more detail by estimating the cost of each assembly. For a bridge, costs for the parapet, deck, pier, abutments, approach slab and slope wall can be estimated individually.

Exhibit 5.4 shows these assemblies: retaining wall, guard wall, roadway, bridge, culvert and storm drainage. The cost shown was calculated by using an approximate estimated cost for each element.

Unit Price

When the design is completed, quantity of each item will be calculated and the cost will be estimated based on unit price of each item.

During the final design, the quantity of each element can be calculated and cost can be estimated by using unit price for each material.

Case Study - STH 34 Bypass Cost

Years ago the client had a preferred alternative to solve the problem. It was not accepted due to the high cost. The VE team made this as an alternative to perform function analysis: Construct three bridges along the existing STH 34 Bypass. The bridges will carry STH 34 Bypass over Canton Street, Maroons Road and Spartan Road. The three crossings will be signalized below the bridges. The three crossings will have a tight diamond interchanges to access the town (see Exhibit 5.2).

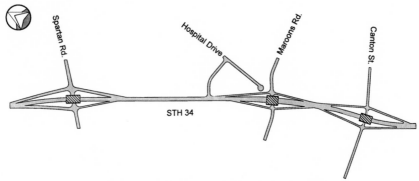

Exhibit 5.2: Alternative to construct three bridges.

For the VE study this alternative was used to estimate cost, allocate functions and function cost. Cost of the proposed STH 34 Bypass improvement case study is shown in Exhibit 5.3.

Item	Quantity	Unit	Unit Cost	Total Cost
Structure	4455	sq.ft	$160	$712,800
Mainline	8500	Lin. ft	$300	$2,550,000
Ramp	1	Each	$98,000	$98,000
Signals	1	Each	$75,000	$75,000
ROW	1	Lsum	$300,000	$300,000
Retaining wall	200	Lin.ft	$1,500	$300,000
Spartan Road improvements	1	Lsum	$150,000	$150,000
MOT	1	Lsum	$100,000	$100,000
Spartan Road interchange				**$4,285,800**
Structure	4455	sq.ft	$160	$712,800
Mainline	0	Lin. ft	$300	$0
Ramp	1	Each	$98,000	$98,000
Signals	1	Each	$75,000	$75,000
Retaining wall	200	Lin.ft	$1,500	$300,000
Maroons Road improvements	1	Lsum	$200,000	$200,000
MOT	1	Lsum	$100,000	$100,000
Maroons Road interchange				**$1,485,800**
Structure	4455	sq.ft	$160	$712,800
Mainline	0	Lin. ft	$300	$0
Ramp	1	Each	$98,000	$98,000
Signals	1	Each	$75,000	$75,000
Retaining wall	200	Lin.ft	$1,500	$300,000
Canton Street improvements	1	Lsum	$150,000	$150,000
MOT	1	Lsum	$100,000	$100,000
Canton Street interchange				**$1,435,800**
Additional turning lanes (3 interchanges)	1	Lsum	$400,000	$400,000
Total				$7,607,400

Exhibit 5.3: Proposed STH 34 Bypass improvements cost

Selection of Elements for VE Study Based on Constraints

During a VE study it is also important to differentiate the cost of the elements that can't be changed. The reason is usually due to stakeholder's constraints. Exhibit 5.4 shows such distribution for a highway project, Rehabilitation of Wheaton Parkway. The total cost of the project was $9.734 million. Cost of the constraints was $4.649 million. The VE team chose to study the elements that cost $5.085 million. This gave the team time to scope the study properly. The team could challenge the constraints if they chose to. However, it would take additional effort and time to support their objections. This left the team to value engineer the replacement of the existing bridge and the construction of new retaining wall as major items. This allowed the VE team to focus on a few elements. By having focused on a small number of elements, the team was able to successfully demonstrate that the bridge didn't have to be replaced at all. This saved construction time, preserved a historical bridge and about 25% of the project cost.

Exhibit 5.4 also lists items that are considered beyond the scope of the study. These items are not studied for the following reasons:

1. Cost is less than 2%

2. Violates constraints

3. Extensive studies were already performed and agreements were made among the stakeholders

4. Does not meet client standards

For example the cost of the pavement and other roadway typical sections is $2.74 million. It is 28 percent of the total cost. However, the clients had done an extensive study to arrive at this typical section. They listed it as a project constraint. The VE Team needed to recognize and respect the project constraints. Similarly, other items that did not meet client standards are also listed under "items not considered" in Exhibit 5.4.

	Item	Subtotal Amount	Total Amount	Items Not Considered	Total Percent
1	Roadway Excavation		$470,000		5
2	Structure Excavation		$210,000		2
3	Retaining Wall		$1,240,000		13
	Concrete	$680,000			
	Reinforcing Steel	$80,000			
	Shoring	$260,000			
	Dampproofing	$10,000			
	Masonry	$140,000			
	Cap Stone	$70,000			
4	Guardwall		$450,000		5
5	Roadway		$2,735,000		28
	Curb and Gutter	$680,000		$680,000	
	Cracking and Sealing	$110,000			
	Asphalt	$1,110,000		$1,110,000	
	Milling	$110,000		$110,000	
	Aggregate Base	$350,000		$350,000	
	Stabilized Soil	$85,000		$85,000	
	Topsoil	$140,000		$140,000	
	Seeding	$50,000		$50,000	
	Sodding	$100,000			
6	Bridge		$892,000		9
	Concrete - Footing	$120,000			
	Concrete - Superstr.	$120,000			
	Concrete - Wingwalls	$160,000			
	Bridge Excavation	$56,000			
	Reinforcing Steel	$60,000			
	Shoring	$70,000			
	Damp proofing	$5,000			
	Masonry	$75,000			
	Cap Stone	$46,000			
	Piles	$180,000			

Exhibit 5.4: Rehabilitation of Wheaton Parkway

Item	Subtotal Amount	Total Amount	Items Not Considered	Total Percent
7 Culvert		$154,000		2
Concrete	$83,000			
Masonry	$26,000			
Cap Stone	$35,000			
Reinforcing Steel	$10,000			
8 Storm Drainage		$1,324,000		14
Headwalls	$54,000			
Pipe Culverts	$468,000			
End Sections	$29,000			
Inlets	$677,000			
Underdrains	$11,000		$11,000	
Riprap	$85,000			
9 Timber Guardrail		$146,000		1
10 Construction Staking		$234,000	$234,000	2
11 Contractor Inspection		$225,000	$225,000	2
12 Schedule/Mobilization		$775,000	$775,000	8
13 Removal Items		$150,000	$150,000	1
14 Erosion Control		$208,000	$208,000	2
15 Signing and Signal		$397,000	$397,000	4
16 Pavement Marking		$58,000	$58,000	1
17 Barricades		$66,000	$66,000	1
Total	**$9,734,000**	**$4,649,000**	**100**	
Cost of Items Studied	**$5,085,000**			

Exhibit 5.4: Rehabilitation of Wheaton Parkway (continued)

Elemental cost

Exhibit 5.4 is a typical cost estimate using material cost such as concrete and reinforcing steel. For function cost analysis it is preferable to estimate cost by elements such as pier, abutments, pavement, etc. The traditional cost estimate, by trade or product, is good for contractors to bid a job. For designers and value engineers the cost of each element is needed to allocate the cost of the element to various functions.

Example - Arch Bridge

Exhibit 5.5 shows a summary version of such elemental cost for an arch bridge. The cost of the double hanger assembly is shown as $700,000 (5%). The original cost estimate was 4% total. The hangers are the critical element that carry the bridge and the traffic. The original cost of 4% was for a single hanger with no redundancy. By having a double hanger assembly (Exhibit 5.6) for an additional 1% of the total cost, the bridge has redundancy. The value of the bridge is substantially increased by increasing the cost by 1%. This is possible by analyzing the cost by elements and functions. Developing an elemental cost estimate is mandatory to any VE study.

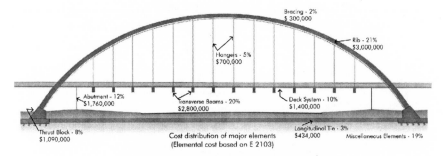

Exhibit 5.5: Arch bridge elemental cost

Exhibit 5.6: Detail of double hanger assembly

ASTM E2103 suggests such elemental cost estimate format facilitates function cost distribution. Exhibit 5.7 shows such distribution. This format is consistent with ASTM E2103 of Bridge UNI-FORMAT II Standard. This format can be used when all elements are defined during design development.

Level 1 Major Group Elements		Level 2 Group Elements		Level 3 Individual Elements
A	Site Work	A10	Utility Relocation		
		A20	Existing Structures Removal		
		A30	Excavation		
		A40	Cofferdam	A4010	Sheeting
				A4020	Under Water Excavation
				A4030	Seal Coat
				A4040	Dewatering
		A50	Embankment		
		A60	Traffic Maintenance		
		A70	Environmental	A7010	Mitigation
				A7020	Protection
		A80	Demolition	A8010	Excavation
				A8020	Removal
B	Substructure	B10	Foundations	B1010	Spread Footings
				B1020	Piles
				B1030	Drilled Shafts, Cap Beams
		B20	Piers	B2010	Cap Beams
				B2020	Columns-single, Multiple
				B2030	Walls-Grade, Crash, Debris
				B2040	Slab Piers
				B2050	Special

Exhibit 5.7: Classification of Bridge Elements

Level 1 Major Group Elements	Level 2 Group Elements		Level 3 Individual Elements	
	B30	Abutments	B3010	Sill Type
			B3020	Spill Through
			B3030	Retaining Wall Type
			B3040	Integral-Semi Integral
			B3050	Vaulted
C Superstructure	C10	Railings	C1010	Traffic
			C1020	Pedestrian
			C1030	Bicycle
	C20	Decks	C2010	Slabs
			C2020	Sidewalks
			C2030	Medians
	C30	Beams	C3010	Stringers
			C3020	Floor Beams
			C3030	Transverse Beams
			C3040	Box Girders
	C40	Special Types	C4010	Tied Arch
			C4020	Suspension
			C4030	Cable Stayed
			C4040	Trusses
	C50	Bearings	C5010	Fixed
			C5020	Expansion
			C5030	Multi-Rotational
	C60	Bracing		
	C70	Movable Mechanism		
D Approach	D10	Wingwalls		
	D20	Retaining Walls	D2010	Cantilever
			D2020	Soldier Pile
			D2030	Sheet Piling
			D2040	MSE Walls
			D2050	Secant Walls
	D30	Approach Slabs		

Exhibit 5.7: Classification of Bridge Elements (continued)

Level 1 Major Group Elements		Level 2 Group Elements		Level 3 Individual Elements	
E	Protection	E10	Expansion Joints	E1010	Open
				E1020	Covered
				E1030	Filled
		E20	Waterproofing		
		E30	Topping/Overlay		
		E40	Drainage	E4010	Scuppers
				E4020	Piping
		E50	Slope Walls		
		E60	Corrosion Control		
		E70	Pier Protection Barriers	E7010	Reinforcement Coating
				E7020	Concrete Admixtures
				E7030	Surface Coating
				E7040	Cathodic
		E80	Approach Barriers		
		E90	General Protection	E9010	Bird Exclusion
F	Services	F10	Signals		
		F20	Signage		
		F30	Lighting		
		F40	Utilities		
		F50	Guard Tower		
		F60	Pavement Marking		
		F70	Inspection and Maintenance	F7010	Deck
				F7020	Below Deck
				F7030	Piers

Exhibit 5.7: Classification of Bridge Elements (continued)

Conclusion

A detailed elemental cost will help the VE team members understand the cost of each element. This effort will be useful in the next step which determines the function cost.

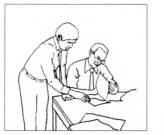

Information Phase—Function Cost

Objective

The objective of this chapter it to allocate cost to each function of the project. Using the project needs, value will be measured by comparing function cost to project needs.

Introduction

When a project is planned, various interest groups influence the size, shape and type of project elements. Various elements are added, modified, enlarged or reduced to fulfill their needs, desires and requirements. Sometimes certain functions cause a major increase in cost. Stakeholders may not be aware of the extent of cost for certain desires. One of the reasons is that the cost of a specific function may be spread among many elements. Allocating the cost of each elements to various functions is the way one can answer the question: "What should it cost?" By allocating costs to various functions, one can better understand the parts or elements. This led to the process of identifying value and mismatch.

Allocation of Element Cost to Functions

- If a element fulfills one function, then the cost of the function is the same as the cost of the element.

- If a element fulfills more than one function, then the cost of the element is apportioned to each function.

Value and Mismatches

When a function or its related part of the element has a high need and the cost is average or below, it has value. When the need is low and the cost is high, it is a mismatch (see Exhibit 6.1). When the cost and need are high, there is an opportunity to improve its value. Improvements can be achieved by lowering the cost of the element or improving its performance.

Need/Performance	Cost	Remarks
High	Average or Low	Value
Low	High	Mismatch
High	High	Opportunity to Improve

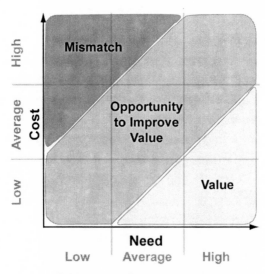

Exhibit 6.1: Value and mismatch

Example – Sidewalk

Consider one element, sidewalk, on an improvement project. The existing condition contains two-way, 12' lanes with two 5' sidewalks (see exhibit 6.2).

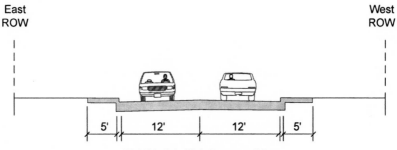

Exhibit 6.2: Existing condition

The As Designed suggests expanding to four-lane highways with two 6' sidewalks. This requires the acquisition of 2' in additional right-of-way (see Exhibit 6.3).

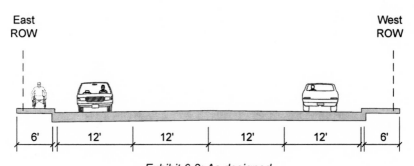

Exhibit 6.3: As designed

Right-of-way (ROW) is needed on the west side to increase the sidewalk width. On the east side, the sidewalk serves an apartment complex, a nursing home and convenience store. On the east side there are busy pedestrian activities. Occasionally, pedestrians sidestep into the roadway to let wheelchair occupants, or elderly people or handicapped people pass by. On the west side of the street the sidewalk is adjacent to a series of automobile spare part industrial facilities. The facility on the west side of the street is accessed mostly by cars. Once in a while someone walks along the sidewalk.

The following table allocates the designed width of the sidewalk to functions:

East Sidewalk

Facilitate Pedestrian Walk	4'	(Exhibit 6.3a)
Facilitate Disabled Walk	2'	(Exhibit 6.3b)
Allow Passing	(additional)	Not Provided

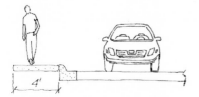 

Exhibit 6.3a: Facilitate Pedestrian   *Exhibit 6.3b: Facilitate Disabled*

West Sidewalk

Facilitate Pedestrian Walk	4'
No Function	2'

The 6' sidewalk on the east side is not wide enough to accommodate handicapped and pedestrians. On the west side there is no need to widen for the handicapped since it is seldom even used by the pedestrians.

Total Cost of Sidewalk			
	Sidewalk Cost	**ROW Cost**	**Total**
East Sidewalk	$54,000	$150,000	$204,000
West Sidewalk	$54,000	$200,000	$254,000

Cost of the sidewalk can be distributed proportionately to these functions using the width ratio.

Function Cost Distribution

Functions	Width	Detail	Function Cost
East Sidewalk			
Facilitate Pedestrian Walk	4'	$204,000 x 4/6 =	$136,000
Facilitate Disabled Walk	2'	$204,000 x 2/6 =	$68,000
Allow Passing	-	-	-
West Sidewalk			
Facilitate Pedestrian Walk	4'	$254,000 x 4/6 =	$169,334
No Function	2'	$254,000 x 2/6 =	$84,667

East Sidewalk

Element	Need	Cost	Remarks
4' sidewalk	High	Average	Value
2' additional	High	Average	Value

West Sidewalk

Element	Need	Cost	Remarks
4' sidewalk	Low	High.	Mismatch
2' additional	No	High	Mismatch

On the east side, observing the volume of traffic an additional two feet will improve value. On the other hand having a sidewalk on the west side just for symmetry, is a mismatch. In this case, during speculations, options such as a reduced minimum width or elimination of west sidewalk, or increase the width of east sidewalk were explored. The use of function cost analysis allows the VE team to understand the importance or impact of the cost compared with its needs. It was proposed to eliminate the west sidewalk and increase the east sidewalk to 8' without needing extra ROW (see Exhibit 6.4). Note that cost is reduced from two 6' sidewalk = 12' to one 8' sidewalk. No cost is needed for extra ROW and the needed functions are satisfied. Value is not necessarily achieved with a wider section, but rather an appropriate section.

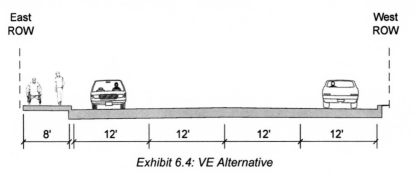

Exhibit 6.4: VE Alternative

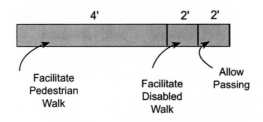

Exhibit 6.4a: East side (8' sidewalk)

Value Index

Technical Function Logic Diagram

One way to measure value is by comparing the cost of the total project to the cost of the critical path functions which are the absolute minimum functions. This can be measured by an index, called "Value Index."

$$\text{Value Index} = \frac{\text{Cost of the Total Project}}{\text{Cost of the Critical Path Functions}}$$

Critical path functions are the bare necessity. By keeping the total cost to as close to the basic needs as possible, it will keep the cost down. A value index under 2.0 is a good target. However, if the cost of the design objectives can be justified, a higher value index may be acceptable. The objective of the calculation of value index is to show how the cost of the proposed solution is distributed.

Customer-Oriented Function Logic Diagram

In a typical transportation project, the distribution for customer-oriented function is close to what is shown below:

Basic Functions	=	20.00%
Assure Dependability Functions	=	30.00%
Assure Convenience Functions	=	25.00%
Satisfy Stakeholders Functions	=	15.00%
Attract Stakeholders Functions	=	10.00%

Based on Muthiah Kasi's work with Thomas J. Snodgrass, Kasi developed the value index for Customer-Oriented Function Logic Diagram

$$\text{Value Index} = \frac{\text{Cost of the Total Project}}{(\text{Cost of the } \textit{Basic Functions} + \text{Cost of } \textit{Assure Dependability} \text{ Functions})}$$

$$\text{Value Index} = \frac{100\%}{(20\% + 30\%)} = 2$$

This relates to the value index the industry uses for the Technical Function Logic Diagram.

Case Study – STH 34 Bypass Function Cost for Technical Function Logic Diagram

The function cost distribution process is not exact. It requires the judgment of a team to distribute the cost using logical reasoning. The solution is to build three bridges over Maroons Road, Spartan Road and Canton Street (see Exhibit 6.5). It includes ramps at each bridge. The ramps at Maroons Road require new ROW. The bridge at Maroons Road will serve the traffic to the existing hospital. The purpose of building a bridge at Canton Street is to attract businesses to a proposed new mall in the area. The need for a bridge across Spartan Road is to serve a proposed high school.

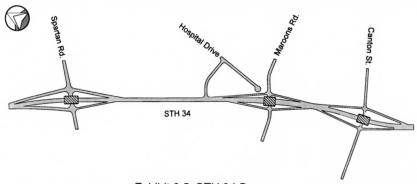

Exhibit 6.5: STH 34 Bypass

Function cost distribution for the Technical Function Logic Diagram is shown in Exhibits 6.6 and 6.6a. The cost of the bridge across Maroons Road is to serve the traffic today. It satisfies the functions *Control Traffic, Separate Traffic* and *Reduce Delays*. The team assumed that the bridge satisfies all three functions equally so the cost of the bridge is allocated equally.

The purpose of the ramps is to maintain local access from the STH 34 Bypass. The cost is allocated to the function *Maintain Access*. The new property acquisition is to accommodate the ramp at Maroons Road and the cost of ROW is allocated to the function *Maintain Access*. The cost of the retaining wall is allocated to the function *Minimize ROW* since it reduces the encroachment of the embankment. Cost of maintenance of traffic (MOT) is to protect the traffic and maintain access during construction. The cost of the bridge across Spartan Road is to serve the future traffic of the proposed high school. The entire cost of this bridge is allocated to the function *Satisfy Long-Term Needs*. The bridge across Canton Street is to encourage business to build a mall. The cost is allocated to the function *Attract Development*. The cost of MOT for Maroons Road was equally divided between the two functions; *Safeguard Residents* and *Maintain Access* and the cost of the three sets of signals is allocated to *Control Traffic*. MOT along Spartan Road and Canton Street is part of the road improvements for today's traffic and the bridge for the future needs. The cost was allocated accordingly. Similar logic was used in the allocation of the roadway cost.

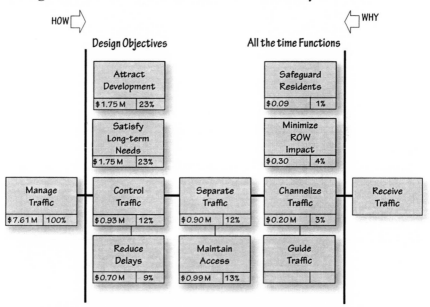

Exhibit 6.6: Technical Function Cost Diagram of STH 34 Bypass

The total cost of the alternative is $7.6 million. Cost of the critical path functions is ($0.93M+$0.90M+$0.20M)=$2.03M. In this case, the value index is $7.61M/$2.03M = 3.75.

The ratio was high due to the design objectives. The team should

	Element Cost	Control Traffic (Basic Function)	Separate Traffic	Channelize Traffic	Reduce Delays	Maintain Access	Safeguard Residents	Minimize ROW Impact	Satisfy Long-Term Needs	Attract Development
		Critical Path			Caused-By Functions		All-the-Time Functions		Design Objectives	
Structure	$712,800	$237,600	$237,600		$237,600					
Ramps	$98,000					$98,000				
Signals	$75,000	$75,000								
ROW	$300,000					$300,000				
Retaining Wall	$300,000							$300,000		
Maroons Road Improvements	$150,000					$150,000				
MOT	$100,000					$50,000	$50,000			
Sub Total - Maroons Road	$1,735,800									
Structure	$712,800								$712,800	
Ramps	$98,000								$98,000	
Signals	$75,000	$75,000								
Retaining Wall	$300,000								$300,000	
Spartan Road Improvements	$200,000					$200,000				
MOT	$100,000					$20,000	$20,000		$60,000	
Sub Total - Spartan Road	$1,485,800									

Exhibit 6.6a: STH 34 Bypass Technical Function Cost

	Element Cost	Critical Path			Caused-By Functions		All-the-Time Functions		Design Objectives	
		Control Traffic (Basic Function)	Separate Traffic	Channelize Traffic	Reduce Delays	Maintain Access	Safeguard Residents	Minimize ROW Impact	Satisfy Long-Term Needs	Attract Development
Structure	$712,800									$712,800
Ramps	$98,000									$98,000
Signals	$75,000	$75,000								
Retaining Wall	$300,000									$300,000
Canton Street Improvements	$150,000					$150,000				
MOT	$100,000					$20,000	$20,000			$60,000
Sub Total - Canton Street	$1,435,800									
Roadway with Maroons Road Bridge Only	$1,400,000	$466,667	$466,667		$466,667					
Additional Road with All Three Bridges	$1,150,000								$575,000	$575,000
Additional Turning Lanes with All Three Interchanges	$400,000		200,000	200,000						
Total Cost	$7,607,400	$929,267	$904,267	$200,000	$704,267	$988,000	$90,000	$300,000	$1,745,800	$1,745,800
		12%	12%	3%	9%	13%	1%	4%	23%	23%
		27%			22%		5%		46%	

Exhibit 6.6a: STH 34 Bypass Technical Function Cost (continued)

question whether the design objective can be modified or changed or find other ways of satisfying them. Later, when and if another alternative is considered for implementation, the value index should be recalculated and compared.

Case Study - STH 34 Bypass Function Cost for Customer-Oriented Function Logic Diagram

Similar logic for cost distribution is applied to the functions of the Customer-Oriented Function Logic Diagram of the same alternative (see Exhibits 6.7 and 6.7a).

The case study function cost distribution is as follows:

Basic Functions	=	12.00%
Assure Dependability Functions	=	16.00% (value is missing)
Assure Convenience Functions	=	22.00%
Satisfy Stakeholders Functions	=	27.00% (mismatch exist)
Attract Stakeholders Functions	=	23.00% (mismatch exist)

The cost for *Attract Stakeholders* cost may go up if a new road is proposed to attract major businesses. The high cost of *Satisfy Stakeholders* functions indicates opportunity to reduce cost without sacrificing value. Later, when and if another alternative is selected for implementation, the cost distribution should be compared.

Value index for the case study is equal to:

$$\text{Value Index} = \frac{100\%}{(12.0\% + 16.0\%)} = 3.57$$

The ratio is high due to the high cost of *Satisfy Stakeholders* and *Attract Stakeholders* functions.

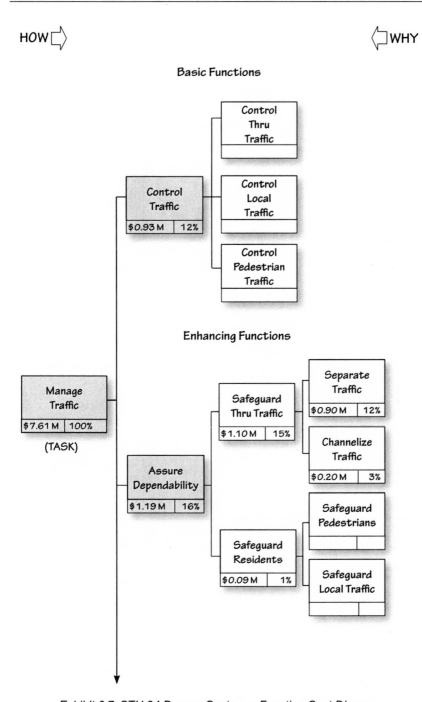

Exhibit 6.7: STH 34 Bypass Customer Function Cost Diagram

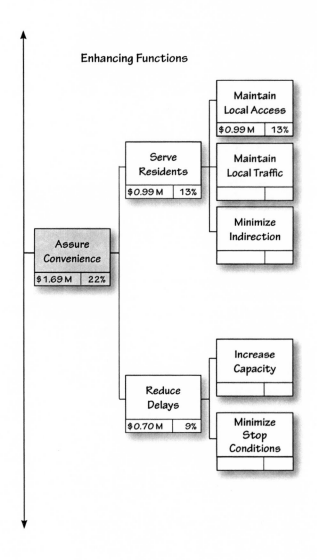

Exhibit 6.7: STH 34 Bypass Customer Function Cost Diagram (continued)

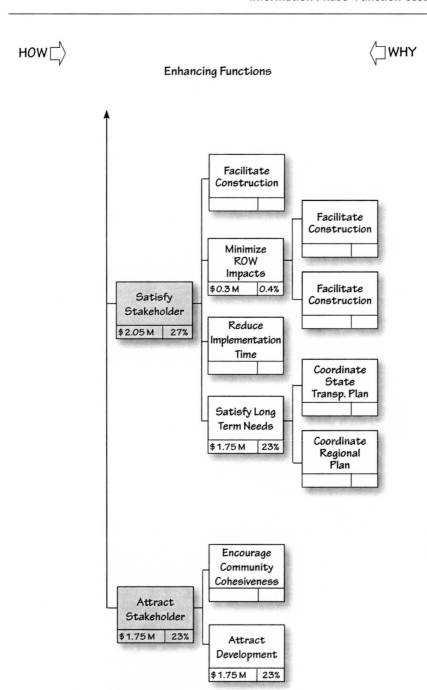

Exhibit 6.7: STH 34 Bypass Customer Function Cost Diagram (continued)

Element	Cost	Basic Functions	Assure Dependability			Assure Convenience		Satisfy Stakeholders		Attract Stakeholders
		Traffic Control	Separate Traffic	Channelize Traffic	Safeguard Residents	Reduce Delays	Maintain Access	Minimize ROW	Satisfy Long-Term Needs	Attract Development
Structure	$712,800	$237,600	$237,600			$237,600				
Ramps	$98,000						$98,000			
Signals	$75,000	$75,000								
ROW	$300,000						$300,000			
Retaining Wall	$300,000							300,000		
Maroons Road Improvements	$150,000						$150,000			
MOT	$100,000				$50,000		$50,000			
Sub Total - Maroons Road	$1,735,800									
Structure	$712,800								$712,800	
Ramps	$98,000								$98,000	
Signals	$75,000	$75,000								
Retaining Wall	$300,000								$300,000	
Spartan Road Improvements	$200,000						$200,000			
MOT	$100,000				$20,000		$20,000		$60,000	
Sub Total - Spartan Road	$1,485,800									

Exhibit 6.7a: STH 34 Bypass Customer Function Cost

Element Cost	Basic Functions	Assure Dependability			Assure Convenience		Satisfy Stakeholders		Attract Stakeholders
	Traffic Control	Separate Traffic	Channelize Traffic	Safeguard Residents	Reduce Delays	Maintain Access	Minimize ROW	Satisfy Long-Term Needs	Attract Development
Structure $712,800									$712,800
Ramps $98,000									$98,000
Signals $75,000	$75,000								
Retaining Wall $300,000									$300,000
Canton Street Improvements $150,000						$150,000			
MOT $100,000				$20,000		$20,000			$60,000
Sub Total - Canton Street $1,435,800									
Roadway with Maroons Road Bridge Only $1,400,000	$466,667	$466,667			$466,667				
Additional Road with All Three Bridges $1,150,000								$575,000	$575,000
Additional Turning Lanes with All Three Interchanges $400,000		$200,000	$200,000						
Total Cost $7,607,400	**$929,267**	**$904,267**	**$200,000**	**$90,000**	**$704,267**	**$988,000**	**$300,000**	**$1,745,800**	**$1,745,800**
	12.22%	11.89%	2.63%	1.18%	9.26%	12.99%	3.94%	22.95%	22.95%
	12.22%	15.70%			22.25%		26.89%		22.95%

Exhibit 6.7a: STH 34 Bypass Customer Function Cost (continued)

Conclusion

Understanding any alternative as it relates to the fulfillment of functions will help the team get ready for the Speculation Phase. Function cost distribution answers the questions "What should it do?" and "What should it cost?"

Value index indicates how close or how far the project is to an effective solution. Either Function Logic Diagram will result in similar value index. After understanding the needs and desires, and how they are not fulfilled, the next step involves creating ideas that can satisfy the stakeholders needs and desires. Equally important is to generate ideas that are based on the functions of the project to arrive at an appropriate solution.

Speculation Phase

Objective

The objective of this chapter is to create as many ideas as possible, through various techniques, that when combined will result in alternatives to the given solution.

Introduction

Creativity is the result of a new combination of thoughts and/or elements based on previously acquired knowledge, exposure and experience. This may be due to imagination, inspiration or illumination. In the Information Phase, we have already learned the needs, desires and constraints of the stakeholders. As a team, we formulated the functions of the project with a logical format using Function Logic Diagrams. A thorough understanding of the issues in the Information Phase is the basis for the imagination process.

When ideas are created without criticism or judgment, the team gets into a rhythm of creativity. This will result in spontaneous stimulation of new and out-of-the-box ideas. If the team goes home at the end of a creative session, ideas will surface at night when in bed or in the morning when in the shower. This is the illumination part of creativity. The creative process is an open-end event. It is effective only if the team is encouraged to imagine.

Rules to Creative Process

The team needs freedom for brainstorming. The following four rules should be observed to ensure maximum creativity.

1. **Criticism is ruled out.** Judgment of all ideas is deferred until after the brainstorming session. You can't be critical and creative at the same time.

2. **Free-wheeling is welcome.** The wilder the ideas, the better. Those apparently wild ideas may turn out to be the solution or they may trigger someone else's thought to suggest a more usable alternative.

3. **Quantity is wanted.** It is easier to pare down a long list of ideas than puff up a short list.

4. **Combination and improvements are sought.** In addition to contributing ideas, try to combine previously proposed ideas into still better ideas and improve upon other ideas.

Guidelines to Creativity

To achieve good results, follow the guidelines outlined here:

Phase 1 – Blast: *Imagine*

- Isolate the basic functions or critical path functions
- Create ideas that will satisfy the basic or critical path functions
- Refine these ideas by adding features that will make it work or accept the concept

- Add features that satisfy one or more of enhancing functions or secondary functions

Phase 2 – Create: Inspire

When the team begins to create, cut them loose. Give them the freedom to list all of their random thoughts.

Phase 3 – Refine: Scream

Listing random ideas is the first step in creative process. To ensure that all possible ideas are listed, test with a matrix, SCREAM: Substitute, Combine, Rearrange, Eliminate, Adapt and Modify. This approach is explained with the following example. Note that the six steps won't be in the order shown above. Sometimes a idea may be modified before it can be combined.

Example - Retaining Wall

A roadway ramp on a 4:1 side slope was proposed to be built under an existing bridge. The embankment was close to an existing steel bent. To protect the existing bearing of the steel bent, a concrete retaining wall was proposed.

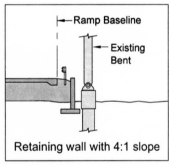

Retaining wall with 4:1 slope

Substitute: Take each idea that is listed. Try to substitute material or shape. If it is concrete, suggest steel, timber, etc. In this case a concrete retaining wall is replaced by steel sheeting.

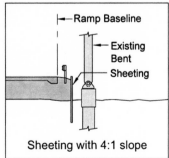

Sheeting with 4:1 slope

Modify: An idea may be enlarged or shortened or the shape may be altered. In this case instead of a 4:1 side slope, a 2:1 side slope is added to the idea. This eliminated the need for the retaining wall.

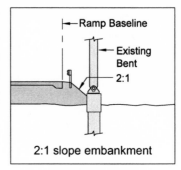

2:1 slope embankment

Combine: Ideas can be easily combined. A 2:1 slope is hard to maintain. An idea detailing a 3:1 slope with a shorter retaining wall is added. This wall can be concrete or substituted with steel sheeting.

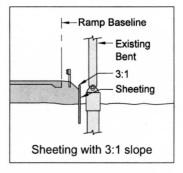

Sheeting with 3:1 slope

Eliminate: The embankment can be eliminated by using structural framing of slab and beams with piles.

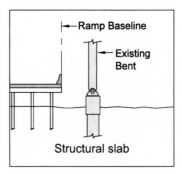

Structural slab

Adopt: Ideas from other case studies can be added. Protecting any elements is accomplished by wrapping the element. A steel casing around the bearing will protect the bearing.

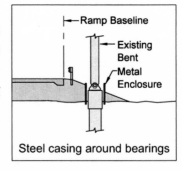

Steel casing around bearings

Rearrange: Rearrangements include moving the elements, reversing, changing places, or placing it somewhere else. In this case, move the concrete wall closer to the bearing. It can also be substituted by a masonry or concrete block wall. Another option is to move the location of the ramp away from the bent.

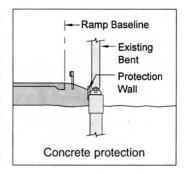

Concrete protection

Using the SCREAM method, the following ideas were generated:

1.	Concrete retaining wall	**Given**
2.	Masonry retaining wall	**S**
3.	Steel sheeting retaining wall	**S**
4.	4:1 side slope	**Given**
5.	2:1 side slope	**M**
6.	3:1 side slope with shorter concrete retaining wall	**C**
7.	3:1 side slope with shorter masonry retaining wall	**S**
8.	3:1 side slope with shorter steel sheeting retaining wall	**S**
9.	Ramp with structural concrete framing on piles	**E**
10.	Ramp with structural steel framing on piles	**S**
11.	Wrap the bearings with sheet metal casing	**A**
12.	Build a small concrete wall on top of the existing pile cap	**R**
13.	Build a small masonry wall on top of the existing pile cap	**S**
14.	Build a small concrete hollow block wall on top of the existing pile cap	**S**
15.	Move the ramp away from the bent to eliminate encroachment	**R**

Phase 4 – Create: *By Functions*

Ideas can be generated by functions. The "How" and "Why" logic critical path will give functions a different level of logical thinking. Using these functions, similar ideas can be generated. The function of the retaining wall is to *Retain Embankment.* Start asking the question "Why?" Why do we retain embankment? The answer is to *Limit Encroachment.* Continue asking the "Why" question. Why do we limit encroachment? The answer is to *Protect Bearing.* Additional ideas can be generated by changing the verb of the function from *limit* to *avoid* or *eliminate.* The above 15 ideas may be generated by function logic approach also as shown below:

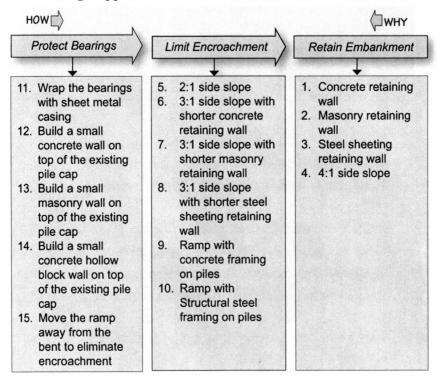

HOW ⇨   ⇦ WHY

Protect Bearings	Limit Encroachment	Retain Embankment
11. Wrap the bearings with sheet metal casing 12. Build a small concrete wall on top of the existing pile cap 13. Build a small masonry wall on top of the existing pile cap 14. Build a small concrete hollow block wall on top of the existing pile cap 15. Move the ramp away from the bent to eliminate encroachment	5. 2:1 side slope 6. 3:1 side slope with shorter concrete retaining wall 7. 3:1 side slope with shorter masonry retaining wall 8. 3:1 side slope with shorter steel sheeting retaining wall 9. Ramp with concrete framing on piles 10. Ramp with Structural steel framing on piles	1. Concrete retaining wall 2. Masonry retaining wall 3. Steel sheeting retaining wall 4. 4:1 side slope

Blocks To Creativity

In addition to understanding how best to be creative, the team members should also be aware of the potential roadblocks. By nature and training, humans are skeptical of change. There are four reasons for such behavior. They are discussed below.

Habitual Blocks
- Continuing to follow a familiar path
- Ignoring ideas that are incompatible
- Conform to custom
- Rely on authority

Perceptual Blocks
- Failure to use all the senses
- Failure to consider the obvious
- Failure to visualize remote relationships
- Failure to distinguish between cause and effect

Cultural Blocks
- Desire to follow standards procedures or customs
- Over confidence
- The drive to be practical
- Belief that creativity is a waste of time
- Too quick to make immediate judgments

Emotional Blocks
- Fear of making a mistake
- Fear of management
- Anxious to succeed quickly

By keeping these four factors in mind, we can be more aware of the potential roadblocks and be better equipped to discourage them and thereby encourage creativity.

Case Study – STH 34 Bypass List of Ideas

Following is the summary of Blast and Create process of creating random ideas for the STH 34 Bypass Project.

1. Erect 2 way stop signs at Maroons Road
2. Erect 2 way stop signs at Spartan Road
3. Erect 2 way stop signs at Canton Street
4. Erect 4 way stop signs at Maroons Road
5. Erect 4 way stop signs at Spartan Road
6. Erect 4 way stop signs at Canton Street
7. Erect signals at Maroons Road
8. Erect signals at Spartan Road
9. Erect signals at Canton Street
10. Prohibit local crossings of STH 34 Bypass at Maroons Road
11. Prohibit local crossings of STH 34 Bypass at Spartan Road
12. Prohibit local crossings of STH 34 Bypass at Canton Street
13. Construct Bridge over Maroons Road
14. Construct Bridge over Spartan Road
15. Construct Bridge over Canton Street
16. Construct bridges over all three crossings (As Given)
17. For idea #16, plan a phased construction
18. Construct Bridge at Maroons Road over STH 34 Bypass
19. Construct Bridge at Spartan Road over STH 34 Bypass
20. Construct Bridge at Canton Street over STH 34 Bypass
21. Construct a roadway tunnel under STH 34 Bypass at Maroons Road
22. Construct a roadway tunnel under STH 34 Bypass at Spartan Road
23. Construct a roadway tunnel under STH 34 Bypass at Canton Street
24. Construct a Pedestrian tunnel under STH 34 Bypass at Maroons Road
25. Construct a Pedestrian tunnel under STH 34 Bypass at Spartan Road

26. Construct a Pedestrian tunnel under STH 34 Bypass at Canton Street

27. Reduce speed on Bypass to 30 mph

28. Permit right-in and right-out only

29. Use roundabouts at Maroons Road only

30. Convert Bypass to a local street

31. For idea #21, use stop signs

32. For idea #21, use signals

33. Construct Pedestrian Bridge at Maroons Road

34. Construct Pedestrian Bridge at Spartan Road

35. Construct Pedestrian Bridge at Canton Street

36. Convert STH 34 Bypass to a local road and build a new Bypass adjacent to the existing Bypass.

37. Convert STH 34 Bypass to a local road and build a new Bypass 1 mile away from the existing Bypass.

38. Utilize retaining wall to eliminate ROW needs.

39. Do nothing

40. Do not build a high school at this location

41. Build only for immediate needs. Improve Maroons Road intersection.

42. Build a single span Bridge

43. Build a two span bridge

44. Build a concrete bridge

45. Build a steel bridge

46. Make provision for future needs

47. Construct a diamond interchange at Maroons Road

48. Construct a diamond interchange at Spartan Road

49. Construct a diamond interchange at Canton Street

50. Construct a tight diamond interchange at Maroons Road

51. Construct a tight diamond interchange at Spartan Road

52. Construct a tight diamond interchange at Canton Street

53. Construct a clover-leaf interchange at Maroons Road

54. Construct a clover-leaf interchange at Spartan Road

55. Construct a clover-leaf interchange at Canton Street

56. Construct a split diamond interchange between Spartan Road and Canton Street
57. Add turning lanes on local streets
58. Build a frontage road

Lateral and Parallel Thinking

When creating ideas, random ideas can be grouped into two types; lateral and parallel. Ideas that are different from others are the result of lateral thinking. Here, the team dares to think outside the box, suggesting nontraditional alternatives to a conventional approach. Ideas enhanced, but similar, are the result of parallel thinking. The team members all tend to go in the same direction and end up getting similar ideas.

For each lateral thinking concept, an alternative can be developed by using some of the parallel thinking ideas. In some cases two lateral thinking concepts, with their appropriate parallel thinking ideas, can be used for an alternative. An example of this approach is shaded in Exhibit 7.8.

Grade separation of the roads to allow all traffic to use the highway is radically different from the idea of prohibiting the local traffic into STH 34 Bypass. Similarly, a grade separation for pedestrians is another example of lateral thinking. With parallel thinking, signals, stop signs, yield signs and signals are similar ideas with minor variations. Notice that the final accepted alternative has a mixture of ideas of both thinking processes.

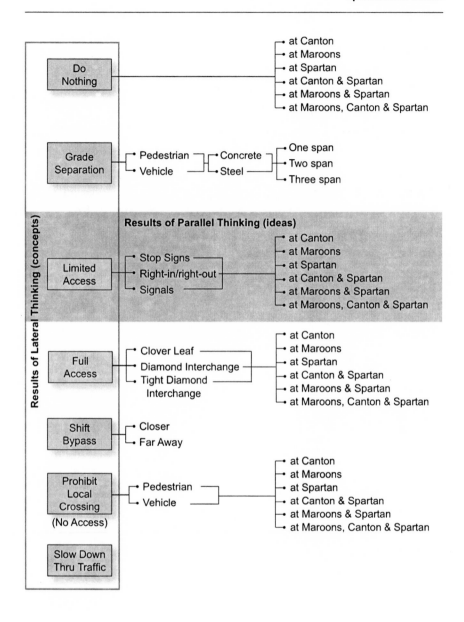

Exhibit 7.8: Lateral and parallel thinking

Conclusion

Various methods to generate ideas were presented in this chapter. They range from micro to macro levels. They also include ideas which are general and specific. Increasing the number of ideas was the driving force of this chapter. The ideas that are generated will be screened for feasibility. This includes a test to see whether ideas violated any constraints. Also, ideas will be grouped under each concept and then they will be converted to alternatives, if needed.

The elimination of ideas is explained in Chapter 8, Evaluation Phase-Screening.

Evaluation Phase–Screening

Objective

The objective of this chapter is to determine which of the ideas developed in the Speculation Phase are worthwhile and then use those ideas to develop alternatives to the As Given solution.

Introduction

Chapter 7 discussed the Speculation Phase where ideas were sought without criticism. The goal of speculation was quantitative rather than qualitative. In the screening process we ask the question "Is this idea worth further consideration or not?" It is essential that the screening process is collaborative and includes the opinions of all VE team members. Even if an idea is felt to be worthwhile by only one member, it should be selected for further consideration. The one member who supports an idea is then the idea's champion and is responsible for any additional work required to develop it further.

Screening Process

In the screening process, the VE team considers whether to select (S) or to reject (R) every idea for future consideration. At this stage in the VE study, the reasoning behind selection or rejection of an idea is based largely on the opinion of the VE team members. However, some ideas clearly violate a stakeholder constraint or are not feasible. These should be rejected.

If an idea is not selected for further consideration, a reason for the rejection must be given and the reason noted as part of the documentation process. As mentioned above, the reason that an idea was rejected is based on the opinion of the VE team at the time of the study. It is possible, in the future, that circumstances may change and that a rejected idea might be resurrected.

Typical reasons for rejecting an idea include: does not satisfy the purpose of the project, is beyond the scope of the project, too costly, impacts project schedule, violates a stakeholder constraint and not feasible. Each VE team will develop their own set of reasons.

The selection of an idea for further consideration does not automatically mean that the idea will become a part of an alternative. It may wind up as good idea that might be applied with all alternative solutions.

Once the screening process is complete, the selected ideas are then matched with the concepts identified in the Function Analysis Phase. This step may result in each concept being an alternative solution or it may result in creating a range of alternatives for a concept.

Ideas that are selected or rejected can be tested for their perceived performance and perceived acceptance. Figure 8.1 shows the four zones for placement of ideas. Zone A includes ideas ranging from neutral to high in perceived acceptance and performance. Most of the ideas fall in this zone. These are traditional ideas that are usually tested and accepted. Ideas in Zone C are perceived to be low in both categories. If a team member sees merit in these ideas the team should encourage that member to be a champion and demonstrate its value. Zone B, in general, are ideas that represent certain stakeholder's desire. They usually conflict with other stakeholder's needs and therefore are viewed as low performing. Zone D, in general, are ideas that represent certain stakeholder's needs. They usually conflict with other stakeholder's desires and therefore are viewed as low in acceptance.

Opinions of the team will be plotted in the screening graph, similar to the one shown in Exhibit 8.1. This screening graph represents the team's opinion only. Its purpose is to show the reasons behind the screening process. In general ideas in Zone A have a higher chance of being selected for further evaluation and ideas in Zone C have a higher chance of being rejected. However, if one member of the team likes an idea in Zone C or in any of the Zones, it will be selected for further evaluation. The one person that supports it is then the champion and is required to develop it for further consideration. Based on the performance and acceptance, ideas will be selected or rejected for further consideration.

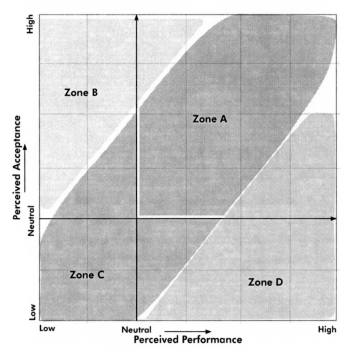

Exhibit 8.1: Screening graph

Enhancing Ideas in Zone C

Inaccurate perceptions can often lead to the rejection of ideas. In such cases, these ideas usually end up at the lower left hand corner of the screening graph, Exhibit 8.2.

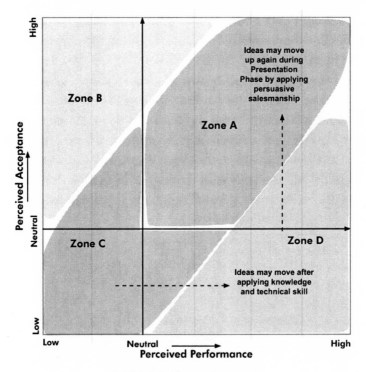

Exhibit 8.2: Screening graph

Our experience shows that most ideas will fall within Zone A. It is the nature of the team behavior to play things safe. It is the responsibility of the team leader to push the team to come up with ideas outside this shaded area. Highest performance ideas are sometime very unpopular, or perceived to be of low performance. The team should propose these types of ideas in the Speculation Phase and develop all the needed reasons in the Development Phase to convince the owner. This will result in a good solution that has great value to the users.

The following steps will help when promoting good ideas:

Step 1: Review ideas that are in the low zone. Even if one team member thinks that it has some merit, investigate.

Step 2: List the potential benefit to the project. If it is high enough, it warrants further development.

Step 3: Technical experts on the subject idea should try to develop it technically and document its strengths and weaknesses. By doing so, the idea may move to the right and end up ranked as high performance with low perceived acceptance. It is up to the team to present their findings with the proper visual data and persuade the decision makers to accept it.

Example - Truss Shortening

The Murray Baker Truss shortening project is a an example that demonstrates how an idea moved from Zone C to Zone A and was eventually accepted as a solution.

This project evolved due to the need to improve an interchange at the north end of the truss. The interchange was built in the late 1950s and had numerous design deviations from current standards, including several sub-standard safety features. Several places within the interchange had accident rates that were more than ten times higher than statewide averages. Thus, the interchange needed improvement, but the proximity of the truss posed a significant site constraint. Shortening the truss was proposed as a solution.

This solution proved to be extremely complex, and was the first known truss shortening of its kind. Determining how much of the truss to remove was crucial to maintaining the balance of the suspended spans. Removal of 180 feet was deemed the optimum length, but significant concerns were raised. There was the possibility of an abrupt energy release with the huge load in the lower cord of 2,200 kips. If the truss was cut and released abruptly, the reverberations would be felt throughout the truss.

Shortening the bridge violated two of the project constraints. One was to maintain traffic on the bridge during construction and the other was to not reconstruct the truss. The idea started as a low acceptance and low performance idea. The Benesch team first worked on the technical aspect of the idea. To alleviate any concerns, Benesch used a custom-designed load transfer device. This device was designed to be placed on the lower cord of the truss to receive 2,200 kips (2.2 million pounds) from the truss member so that it could be cut without any load on it. By answering all the technical questions, the idea moved to a high performance and low acceptance zone. The project manager used his technical and persuasive skills to demonstrate the long-term benefit to the clients and stakeholders and moved the idea to a high performance and high acceptance solution.

Exhibit 8.4a: Original truss

Exhibit 8.4b: Truss after shortening

Assembling of Alternatives

Selected ideas are used to develop alternatives to satisfy the project requirements.

Step 1: Pick a concept from the list of lateral thinking

Step 2: List all the accepted ideas that are related to this concept

Step 3: Assemble an alternative

Step 4: List the functions that will be satisfied by this alternative. This will help add or subtract elements from the alternative.

Case Study – STH 34 Bypass Screening of Ideas

The ideas generated in the Speculation Phase are screened for further evaluation. Ideas are either rejected or selected for further evaluation and development.

Possible Ideas	Screening Results
1. Erect 2 way stop signs at Maroons Road	S
2. Erect 2 way stop signs at Spartan Road	S
3. Erect 2 way stop signs at Canton Street	S
4. Erect 4 way stop signs at Maroons Road	R1
5. Erect 4 way stop signs at Spartan Road	R1
6. Erect 4 way stop signs at Canton Street	R1
7. Erect signals at Maroons Road	S
8. Erect signals at Spartan Road	S
9. Erect signals at Canton Street	S
10. Prohibit local crossings of STH 34 Bypass at Maroons Road	R3
11. Prohibit local crossings of STH 34 Bypass at Spartan Road	R3
12. Prohibit local crossings of STH 34 Bypass at Canton Street	R3
13. Construct Bridge over Maroons Road	S
14. Construct Bridge over Spartan Road	S
15. Construct Bridge over Canton Street	S

16.	Construct bridges over all three crossings (As Proposed)	S
17.	For idea #16, plan a phased construction	S
18.	Construct Bridge at Maroons Road over STH 34 Bypass	S
19.	Construct Bridge at Spartan Road over STH 34 Bypass	S
20.	Construct Bridge at Canton Street over STH 34 Bypass	S
21.	Construct a roadway tunnel under STH 34 Bypass at Maroons Road	S
22.	Construct a roadway tunnel under STH 34 Bypass at Spartan Road	S
23.	Construct a roadway tunnel under STH 34 Bypass at Canton Street	S
24.	Construct a Pedestrian tunnel under STH 34 Bypass at Maroons Road	R2
25.	Construct a Pedestrian tunnel under STH 34 Bypass at Spartan Road	R2
26.	Construct a Pedestrian tunnel under STH 34 Bypass at Canton Street	R2
27.	Reduce speed on Bypass to 30 mph	R1
28.	Permit right-in and right-out only	R2
29.	Use roundabouts at Maroons Road only	R1
30.	Convert Bypass to a local street	R1
31.	For idea #21, use stop signs	S
32.	For idea #21, use signals	S
33.	Construct Pedestrian Bridge at Maroons Road	R1
34.	Construct Pedestrian Bridge at Spartan Road	R1
35.	Construct Pedestrian Bridge at Canton Street	R1
36.	Convert STH 34 Bypass to a local road and build a new Bypass adjacent to the existing Bypass	S
37.	Convert STH 34 Bypass to a local road and build a new Bypass two miles away from the existing Bypass	S
38.	Utilize retaining wall for idea #50-52 to eliminate ROW needs.	DS

39.	Do nothing	R3
40.	Do not build a high school at this location	R4
41.	Build only for immediate needs / Improve Maroons Road intersection	S
42.	Build a single span bridge	DS
43.	Build a two span bridge	DS
44.	Build a concrete bridge	DS
45.	Build a steel bridge	DS
46.	Make provision for future needs	DS
47.	Construct a diamond interchange at Maroons Road	S
48.	Construct a diamond interchange at Spartan Road	S
49.	Construct a diamond interchange at Canton Street	S
50.	Construct a tight diamond interchange at Maroons Road	S
51.	Construct a tight diamond interchange at Spartan Road	S
52.	Construct a tight diamond interchange at Canton Street	S
53.	Construct a clover-leaf interchange at Maroons Road	R2
54.	Construct a clover-leaf interchange at Spartan Road	R2
55.	Construct a clover-leaf interchange at Canton Street	R2
56.	Construct a split diamond interchange between Spartan Road and Canton Street	S
57.	Add turning lanes on local streets	S
58.	Build a frontage road	S

Key	
S	Selected
R1	Rejected - Violates Constraint
R2	Rejected - Inconvenience
R3	Rejected - Does not satisfy the purpose of the project
R4	Rejected - Beyond the scope of the project
DS	Design Suggestion

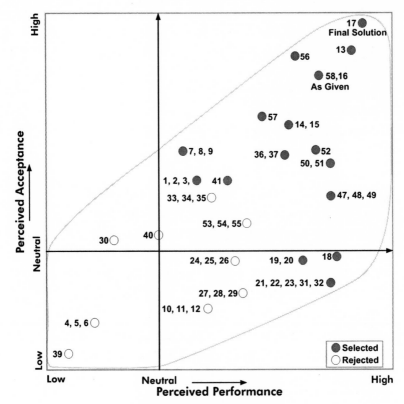

Exhibit 8.3: Screening graph

Look at some of the ideas for the case study STH 34 Bypass improvements (see Exhibit 8.3). Erecting 2-way stop signs along local roads (Ideas 1, 2, 3) improves safety. Thru traffic will accept this idea, but local traffic will not be happy about it. Erecting 4-way stop signs (Ideas 4, 5, 6) will have an impact on thru traffic safety, and speed. Those ideas are placed closer to low performance and low acceptance zone and are then rejected. Constructing three bridges (Idea 16) improves safety but costs more. It has higher performance and high acceptance. If these bridges are built in phases on an as needed basis (Idea 17), it has even higher acceptance. Similarly, other ideas are placed in the chart.

Assembling of Alternatives

Various options were speculated, and more than 10 options were proposed. During part of the initial screening some options were dropped due to high cost, violation of constraints or because they did not satisfy the basic functions of the project.

Four alternatives were developed from the list of lateral thinking ideas. They were then modified or changed to make them more viable. An example of the evolution of one alternative (Alternative 7A) is presented here. A similar approach was used to develop other alternatives.

Assembling of Alternative 7A

Step 1: Pick a concept from the list of lateral thinking: Grade separation vehicular

Step 2: List all the accepted ideas to this concept

- Prohibit local crossing related ideas 10-12
- Bridge related ideas 13 thru 23, 42-45
- Interchange related ideas 47-52
- Access types ideas 57 and 58

Step 3: Assemble an alternative

- Use existing alignment for thru traffic
- Build a Frontage road for local traffic (Idea 58)
- Prohibit Local crossings (Ideas 10-12)
- Construct Bridges over three crossings (Idea 16)
- Construct Split diamond interchange between Canton Street and Spartan Road

Step 4: List the functions that will be satisfied by this alternative. This will help add or subtract elements from the alternative. Alternative 7 was similar to 7A; however it included three full interchanges. To minimize ROW, Alternative 7 is changed to

7A with a split diamond at the ends.

After the initial screening, the following alternatives were selected for further evaluation.

Alternative 3A: Install stop and go signals at intersections with Maroons Road and Spartan Road, as well as channels to separate oncoming traffic. Add storage lanes for left-turn vehicles.

Alternative 4A: Construct a bridge at Maroons Road and install stop-lights at Spartan Road. Add a frontage road between Canton Street and Maroons Road.

Alternative 7A: Construct three bridges over Canton Street, Maroons Road and Spartan Road. Build the STH 34 Bypass and bridges along the existing alignment. Construct a new local road or frontage road adjacent to the existing STH 34 Bypass. Build partial ramps at the beginning and at the end as shown in the Exhibit 8.4

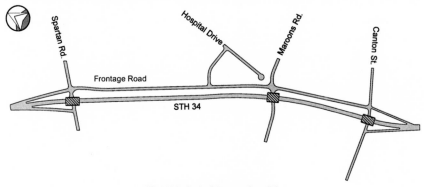

Exhibit 8.4: Alternative 7A

Alternative 9A: Build a new bypass highway about two miles east of the STH 34 Bypass.

In addition to the selected alternatives, include "Do Nothing" or "Existing Condition" as a mandatory alternative to evaluate. Most governmental agencies require the planning comparison to include the impact of "Do Nothing" scenario. From a VE point of view, it should be included as a bench mark or reference. At the end we need to know how much benefit an alternative achieves in both performance and acceptance, and at what cost.

The screening process ends with a list of alternatives that are perceived to be qualified to solve the problem. The next step is to test them for performance, acceptance and cost. The ideas that are not included in the alternatives are available to be incorporated into the alternatives after the rating and ranking.

Conclusion

The screening process narrows down the ideas that can be used to assemble alternatives. Plotting the results of the screening process on a graph can visually guide the team on how each idea is perceived. If the team takes a conservative position, they can assemble a set of alternatives from Zone A. If they feel strongly about an idea in Zone C, they should be prepared to put forth the extra effort to demonstrate its performance value and campaign for its acceptance. Plotting the results is an optional step. The next step is to take the alternatives through the ranking process which will be discussed in Chapter 9 Evaluation Phase–Ranking.

Evaluation Phase–Ranking

Objective

The objective of evaluation is to reduce the quantity of ideas and then qualify them as potential alternatives by rating and ranking the selected alternatives for development.

Introduction

During the Evaluation Phase, alternatives and ideas are measured for their worth. There are three major components considered during this phase: performance, acceptance and cost.

Throughout this process, answers will be revealed to questions such as:

"Will it work?"
"Will it be acceptable?"
"Can we afford it?

A favorable alternative will have a high rating in both performance and acceptance. In other words, the alternative will maintain full functionality while simultaneously satisfying the stakeholder's needs and desires.

Performance is the measurement of how well an alternative meets the functions of the project. For example, an alternative for a highway will have capacity, access and safety related issues as performance criteria. Performance criteria are measurable. For example if the highway needs to handle 24,000 vehicles per day (vpd) the capacity for each alternative can be calculated.

Acceptance is an indication of how well it will be perceived as a good or bad alternative to the stakeholders. For example, an alternative for a highway project will affect the land usage. How will it improve the land value? The acceptance criteria are subjective and consider the perception or feelings of the stakeholders.

The value of an alternative is achieved when the alternative can reliably perform all the needed functions, is acceptable to the stakeholders and comes at a reasonable cost to the owner.

It is important that performance, acceptance and cost criteria are not combined for evaluation purposes. They are conflicting and must be evaluated separately. Then the results are compared with various degrees of importance.

Evaluation Phase Process

Accurately judging the alternatives requires an intricate process. Part of the process is subjective and part is objective. Judging is done sequentially in three independent steps: Performance Rating, Acceptance Rating and Cost Rating. To begin, each alternative is named (Alternative 1, 2, 3, etc.), and then they are selected one at a time for evaluation. This process occurs in two phases. The first phase includes the following steps for performance and acceptance ratings:

Step 1: Develop criteria
First, consider all of the different criteria that should be a factor within this evaluation. Does the project have a strong community opinion? Is the owner trying to meet a tight schedule? Are there political factors that have a role in this

decision? All of the different criteria need to be discussed with the group, and then documented.

Criteria for performance and acceptance are derived from the Function Logic Diagram. All alternatives that are selected for evaluation must satisfy the basic functions. It is therefore a safe assumption that the alternatives, in most cases, need not be evaluated for basic functions.

Step 2: Separate performance and acceptance criteria

Once all the criteria needing consideration in the evaluation process are listed, they need to be separated into performance or acceptance categories. For example, project functionality ideas would fall under performance, while stakeholder buy-in would be acceptance.

The criteria for performance ratings are usually derived from the enhancing functions *Assure Dependability* and *Assure Convenience.* The criteria for acceptance ratings are usually derived from the four enhancing functions; *Assure Dependability, Assure Convenience, Satisfy Users* and *Attract Users.*

Step 3: Rank and assign weights of importance to the criteria

Each criterion is compared against other criteria to test how important each one is in relation to others. This will result in relative ranking of each criterion. The weight of each criterion is based on a scale from 1 to 10, 10 being the most important. The number 1 ranked criterion will be assigned 10 and the next ranked criterion with 10 or less. If the highest ranked criterion is very important, it is not unusual for the next criterion to have a weight of importance 7 or 8.

Step 4: Rate each alternative against the criteria

The next step is to start rating the alternative for each criterion. The scale for this rating is 1 to 5 with 5 being excellent. There are two ways ratings are done, absolute and relative. In relative approach examine each given alternative how it satisfies the criteria. The alternative that satisfies the most gets a 5, the next one gets a 4 or 3 based on its performance compared to the leading alternative. In the absolute approach, the performance of each alternative is tested against a desired level. The desired level should be determined first. For example, if the desired capacity is 24,000 vpd, one alternative has 25,000 vpd it gets a 5. If the second alternative has a capacity of 20,000 vpd, it may get a 3 and so on. The recommended approach is the latter one. The absolute approach truly measures the performance or acceptance of each individual alternative.

Step 5: Compute weighted rating

Multiply the weight of importance of the criterion with the rating of an alternative for the criterion.

Step 6: Compute average weighted rating

Sum the total weight of importance of all criteria. For each alternative add the weighted ratings against all the criteria. Divide this sum by the total weight of importance. This number is the average weighted rating of the alternative.

Cost rating is very subjective. The limits are usually set by the decision makers. It is very critical that the decision makers or the team set proper limits that have a great impact on ratings. Cost for rating 1 and 5 should be set first. The variation between 1 and 5 can be linear or it can be curved. In most cases it is sufficient to assume a linear variation.

Sensitivity Analysis

Phase two of the evaluation process requires that the value engineering team measure the impact of the three ratings. It is very difficult to take a position and say whether performance, acceptance or cost has more or less importance. The best way is to approach this is to explore "what if" scenarios by assigning a different weight of importance to each rating. Ask what if one rating is better than the other two and by how much. Assign weight of importance to performance (P), acceptance (A) and cost (C) ratings. Start with P=1, A=1 and

C=1 and compute the value indicator. Value indicator is the average weighted rating summary of performance, acceptance and cost ratings.

Experience has shown that in various scenarios certain alternatives always end up ranked in the top three. Practically speaking, such an alternative has a better chance of acceptance across the board and in most cases may be within the reach of budget constraints and acceptability.

Case Study – STH 34 Bypass Ranking

As discussed earlier, all alternatives are assumed to satisfy the basic functions. It is therefore not necessary to select criteria from the basic function group of the Function Logic Diagram. In the case study, all alternatives can *Manage Traffic*, the basic function of the STH 34 Bypass. The difference in the selected alternatives is how well they assure dependability, assure convenience, satisfy stakeholders and attract stakeholders. Naturally the criteria will come from these four groups of enhancing functions.

Exhibit 9.1 shows the Function Logic Diagram of the STH 34 Bypass improvements. The criteria selected are shown in the Function Logic Diagram.

Functions that address safety, access and movements are listed as performance criteria. Planning compatibility, construction time, construction difficulty, and impacts to land and development are listed as acceptance criteria. Selection and sorting of criteria are very critical in the selection of an alternative. By overemphasizing certain functions, the selection of the preferred alternative may be narrowed down without proper consideration to multiple needs and desires.

There are three safety criteria, one each for thru traffic, local traffic and pedestrians. There are three criteria for local access and local traffic movements. The team members representing local community suggested three criteria for their convenience; easy construction for the local contractors, less ROW impact to the properties and quicker implementation of the construction plan. The local business people wanted the construction to attract businesses to the community. The following criteria were a result of the state and the county insistence on a plan that would satisfy long-term needs.

Safety and access are considered as performance criteria. These are technical issues and measurable for their performance. Functions representing these criteria are noted in the Function Logic Diagram in Exhibit 9.1 as P series for performance and A series for Acceptance. The following are grouped under Performance Criteria.

1. Through Traffic Safety
2. Local Traffic Safety
3. Access from STH 34
4. Thru Traffic Movement
5. Local Traffic Movement
6. Access to STH 34
7. Pedestrian Safety

The other criteria discussed above are subjective and not measurable. They are grouped under Acceptance Criteria.

1. Compatibility with Long Term
2. Compatibility with Regional Plan
3. Effect of Future Development
4. ROW Impact
5. Constructability
6. Public Acceptability*
7. Time of Implementation

*This criteria was added by the VE team to measure overall acceptance by the public since this is a controversial issue to them.

Criteria are often chosen by the team members representing the stakeholders' interests. Sometimes they favor one area such as safety, convenience, business opportunity or long term planning. By testing them with a Function Logic Diagram the team can balance them to meet the project needs. The criteria for the case study are presented as is. Note that access was represented with two performance criteria. This may result in slightly favoring alternative with local access emphasis. The team leader in this case didn't see this as an issue. The point is the selection of criteria is not an exact science. It is subjective and it represents the team's definition of value for the project.

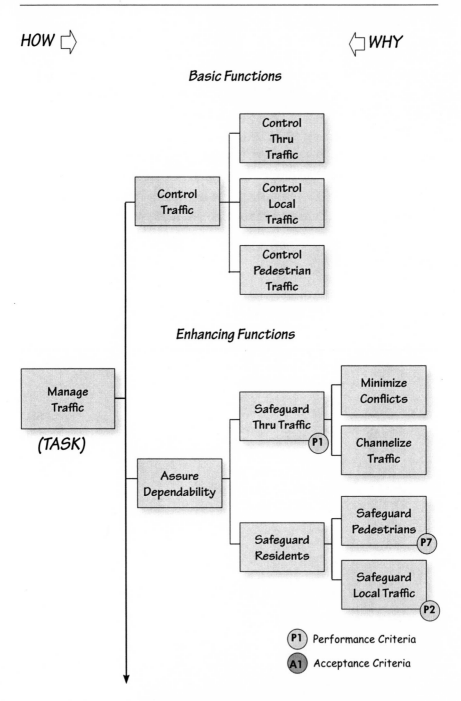

HOW ➡

⬅ WHY

Basic Functions

Enhancing Functions

Control Thru Traffic

Control Local Traffic

Control Pedestrian Traffic

Control Traffic

Manage Traffic

(TASK)

Assure Dependability

Safeguard Thru Traffic · **P1**

Minimize Conflicts

Channelize Traffic

Safeguard Residents

Safeguard Pedestrians · **P7**

Safeguard Local Traffic · **P2**

P1 Performance Criteria

A1 Acceptance Criteria

Exhibit 9.1 Function Logic Diagram

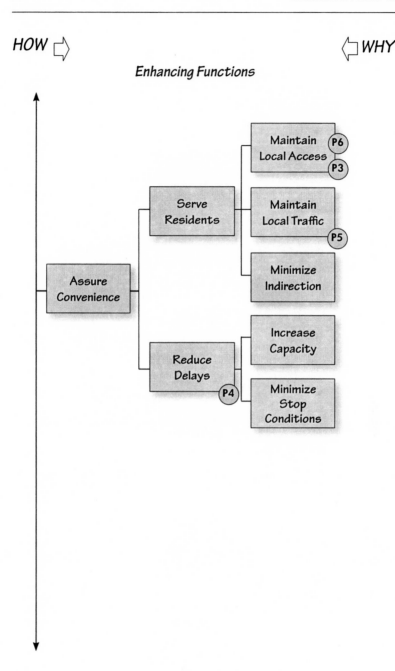

Exhibit 9.1 Function Logic Diagram (continued)

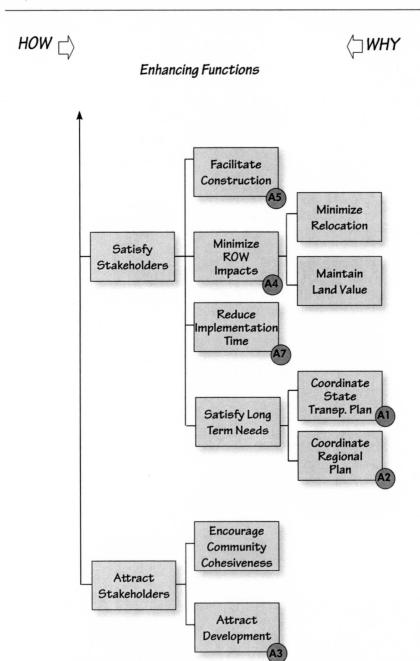

HOW ⇨   ⇦ WHY

Enhancing Functions

Exhibit 9.1 Function Logic Diagram (continued)

Ranking of Performance Criteria

The performance criteria were compared to each other and then ranked in order of importance and assigned a weight of importance on a scale of 1 to 10, see Exhibit 9.2. Each criterion was compared against the other criteria for its relative importance.

The team asked questions, such as: "Is criterion #1, *Thru Traffic Safety*, more important then criteria #2, *Local Traffic Safety*? If the answer is "yes," place a "1". If the answer is "no" place a "0". The answer was "yes" so a "1" was placed in the first column of *Thru Traffic Safety*. Similarly a "1" for criteria #1 was placed under criteria #3 and #4. Note that each criterion is compared against itself first, and given a 1. Place a 1 diagonally for each criterion as shown in Exhibit 9.3.

Criteria Ranking Chart (Performance)							
Criteria	**1. Thru Traffic Safety**	**2. Local Traffic Safety**	**3. Access from STH 34**	**4. Thru traffic Movement**	**5. Local Traffic Movement**	**6. Access to STH 34**	**7. Pedestrian Safety**
1. Thru Traffic Safety	1	1	1	1	1	1	1
2. Local Traffic Safety		1	1	1	1	1	1
3. Access from STH 34			1	0	0	1	0
4. Thru traffic Movement				1	1	1	0
5. Local Traffic Movement					1	1	0
6. Access to STH 34						1	0
7. Pedestrian Safety							1

Exhibit 9.2 Performance criteria ranking chart

At each diagonal, the horizontal row and vertical column will be opposite numbers. The rest of the criteria were compared similarly. When the numbers were added vertically, the ranking of criteria was obtained. If there is a breakdown in logic, two criteria may have the same ranking. For example, there may be two 3's and either a 2 or 4 may be missing. Recheck the logic to correct this error.

Criteria Ranking Chart (Performance)							
Criteria	1. Thru Traffic Safety	2. Local Traffic Safety	3. Access from STH 34	4. Thru traffic Movement	5. Local Traffic Movement	6. Access to STH 34	7. Pedestrian Safety
1. Thru Traffic Safety	1	1	1	1	1	1	1
2. Local Traffic Safety	0	1	1	1	1	1	1
3. Access from STH 34	0	0	1	0	0	1	0
4. Thru traffic Movement	0	0	1	1	1	1	0
5. Local Traffic Movement	0	0	1	0	1	1	0
6. Access to STH 34	0	0	0	0	0	1	0
7. Pedestrian Safety	0	0	1	1	1	1	1
Rank	1	2	6	4	5	7	3
Weight of Importance							

Exhibit 9.3 Performance criteria ranking chart

Assigning Weight of Importance

In assigning the weight of importance, the number one ranked criterion, *Thru Traffic Safety*, was given a 10. The second most important criterion must be given a weight of importance of 10 or less. In this case, the weight of importance given to *Local Traffic Safety* was a 10. The third most important criterion, *Pedestrian Safety*, must be given a weight of importance of 10 or less. In this case, it was given a weight of 9 (see Exhibit 9.4). When weight of importance is 5 or less, the impact of the criteria on the rating diminishes.

Criteria Ranking Chart (Performance)							
Criteria	1. Thru Traffic Safety	2. Local Traffic Safety	3. Access from STH 34	4. Thru traffic Movement	5. Local Traffic Movement	6. Access to STH 34	7. Pedestrian Safety
1. Thru Traffic Safety	1	1	1	1	1	1	1
2. Local Traffic Safety	0	1	1	1	1	1	1
3. Access from STH 34	0	0	1	0	0	1	0
4. Thru traffic Movement	0	0	1	1	1	1	0
5. Local Traffic Movement	0	0	1	0	1	1	0
6. Access to STH 34	0	0	0	0	0	1	0
7. Pedestrian Safety	0	0	1	1	1	1	1
Rank	1	2	6	4	5	7	3
Weight of Importance	10	10	7	8	7	6	9

Exhibit 9.4 Performance criteria ranking chart

Ranking of Acceptance Criteria

The acceptance criteria were similarly compared, ranked and weighted, see Exhibit 9.5. The next step uses these weighted criteria to rate the alternatives for performance and acceptance.

Criteria Ranking Chart (Acceptance)							
Criteria	1. Compatibility w/ Long Term	2. Compatibility w/ Reg. Term	3. Effect of Future Development	4. ROW Impact	5. Constructability	6. Public Acceptability	7. Time of Implementation
1. Compatibility w/Long Term	1	1	1	1	1	1	1
2. Compatibility w/Reg. Term	0	1	1	1	1	1	1
3. Effect of Future Development	0	0	1	1	1	0	0
4. ROW Impact	0	0	0	1	1	0	0
5. Constructability	0	0	0	0	1	0	0
6. Public Acceptability	0	0	1	1	1	1	1
7. Time of Implementation	0	0	1	1	1	0	1
Rank	1	2	5	6	7	3	4
Weight of Importance	10	9	7	6	5	9	8

Exhibit 9.5

Performance Evaluation of Alternatives

Using the weighted performance criteria discussed above, the five alternatives were evaluated based on how well they satisfied each of the performance criteria, see Exhibit 9.6. Each alternative is rated on how well it performs with respect to the criteria on a scale of 5 to 0, where 5 is excellent, 4 is very good, 3 is good, 2 is satisfactory, 1 is poor and 0 is unsatisfactory. By multiplying the ratings by the weight of the criteria and adding the products, a total score for each alternative is found. This score divided by the sum of the weight of importance is the average rating of the alternative, which is used to select the optimum solution. Exhibit 9.7 shows a similar matrix for acceptance.

Procedure for Performance Rating

Step 1: List the criteria and the weight of importance from Exhibit 9.4.

Step 2: Sum the weight of importance.

Step 3: For each criterion, test each alternative to measure how it satisfies the criterion. For example take the criterion, *Thru Traffic Safety*. A new bypass away from the town (Alternative 9A) will give the excellent safety condition to the *Thru Traffic Safety*. Therefore it was rated as 5 (Excellent). By placing a signal (Alternative 3A), the fast moving thru traffic may experience unexpected stop conditions. It is rated 1.5 (Between poor and satisfactory). Similarly all alternatives were given ratings.

Step 4: Each rating is then multiplied by the weight of importance of the criteria. For example, Alternative 3A rating of *Thru Traffic Safety* criterion, the weighted rating is 1.5 * 10 = 15.

Step 5: Sum the weighted ratings for each alternative. For Alternative 3A, the total weighted rating is 120.5.

Step 6: Compute average weighted rating for each alternative. For alternative 3A the average weighted rating is 120.5/54 = 2.11. This number can be compared with the rating scale of 1 (poor) to 5 (Excellent). Alternative 3A is rated as a satisfactory solution.

Similarly, Exhibit 9.7 shows the acceptance rating of the five alternatives.

Guidelines for the Rating Process

The ratings in a team environment should observe the following:

- Limit the number of criteria to 7. The ideal number is 5 or 6.

- Check that no one area, such as capacity or access or safety, has extremely high weights of importance when compared with other important areas.

- The rating should be absolute, not relative by comparison to other alternatives. If all alternatives are poor, they should be rated as 1. If they are all equally good, they should be rated as 3.

- If the ratings on an alternative are widely different, investigate. The members who rated low and high should discuss, resolve and rate again. If the wide difference still exists, the team leader drops the high and low and averages the rest of the members' ratings.

- The members should agree on scale definitions. For example if the criterion is *Pedestrian Safety*, agree on a scale.

  1 = no crossing provisions

  2 = stop signs with crossing striping

  3 = signals

  4 = grade separated

  5 = separate pedestrian bridges

- If an alternative has a high total weighted rating and yet it has a poor (1) rating in one criterion, it should be investigated. The leading alternative cannot have any weakness and still be considered for implementation. Either the alternative should be dropped from further consideration or should be improved.

Excellent = 5 Very Good = 4 Good = 3 Satisfactory = 2 Poor = 1 Unsatisfactory = 0	Weight of Importance	Do Nothing		Alt. 3A		Alt. 4A		Alt. 7A		Alt. 9A	
		Rating	Weighted Rating	Rating	Weighted Rating	Rating	Weighted Rating	Rating	Weighted Rating	Rating	Weighted Rating
Criteria	(1-10)	(1-5)		(1-5)		(1-5)		(1-5)		(1-5)	
1. Thru Traffic Safety	10	3.0	30.0	1.5	15.0	2.0	20.0	4.5	45.0	5.0	50.0
2. Local Traffic Safety	10	1.0	10.0	2.0	20.0	2.5	25.0	4.5	45.0	4.0	40.0
3. Access from STH 34	7	5.0	35.0	3.5	24.5	2.5	17.5	3.0	21.0	3.5	24.5
4. Thru Traffic Movement	8	4.0	32.0	1.0	8.0	2.0	16.0	4.5	36.0	5.0	40.0
5. Local Traffic Movement	7	2.0	14.0	2.0	14.0	3.0	21.0	4.5	31.5	4.5	31.5
6. Access to STH 34	6	5.0	30.0	3.5	21.0	2.5	15.0	3.0	18.0	3.0	18.0
7. Pedestrian Safety	9	1.0	9.0	2.0	18.0	2.5	22.5	4.5	40.5	3.0	27.0
Total Weighted Rating	57		160.0		120.5		137.0		237.0		231.0
Average Weighted Rating		2.81		2.11		2.40		4.16		4.05	

Exhibit 9.6: Performance Rating of STH 34 Bypass alternatives

Excellent = 5 Very Good = 4 Good = 3 Satisfactory = 2 Poor = 1 Unsatisfactory = 0	Weight of Importance	Do Nothing		Alt. 3A		Alt. 4A		Alt. 7A		Alt. 9A	
		Rating	Weighted Rating	Rating	Weighted Rating	Rating	Weighted Rating	Rating	Weighted Rating	Rating	Weighted Rating
Criteria	(1-10)	(1-5)		(1-5)		(1-5)		(1-5)		(1-5)	
1. Compatibility w/Long Term	10	4.0	40.0	1.0	10.0	2.0	20.0	4.5	45.0	5.0	50.0
2. Compatibility w/Reg. Plan	9	4.0	36.0	1.0	9.0	2.0	18.0	4.0	36.0	5.0	45.0
3. Effect of Future Development	7	3.0	21.0	2.0	14.0	3.0	21.0	4.0	28.0	5.0	35.0
4. ROW Impact	6	5.0	30.0	4.5	27.0	3.5	21.0	3.5	21.0	1.0	6.0
5. Constructability	5	5.0	25.0	4.0	20.0	3.0	15.0	2.0	10.0	5.0	25.0
6. Public Acceptability	9	0.0	0.0	3.0	27.0	3.0	27.0	4.5	40.5	5.0	45.0
7. Time of Implementation	8	5.0	40.0	4.0	32.0	3.0	24.0	2.5	20.0	1.0	8.0
Total Weighted Rating	54		192.0		139.0		146.0		200.5		214.0
Average Weighted Rating		3.56		2.57		2.70		3.71		3.96	

Exhibit 9.7: Acceptance Rating of STH 34 Bypass alternatives

Cost Evaluation of Project Study Alternatives

The cost rating is calculated in two steps. Step 1 aims to establish upper and lower limits of the cost rating scale. It was agreed to assume $2 million or less as excellent (5) and $14 million as unacceptable (0). A straight line interpolation is selected to establish intermediate points.

Step 2 aims to mark all the alternatives with their respective cost along this line. Ratings of these alternatives can then be read from this chart (Exhibits 9.8) and on the accompanying graph below (Exhibit 9.8a).

Cost Rating		
Alternative	Cost in Millions	Rating
3A	$1.5	5
4A	$2.0	5
7A	$7.4	2.8
9A	$11.5	1.0

Exhibit 9.8: Cost Rating of STH 34 Bypass alternatives

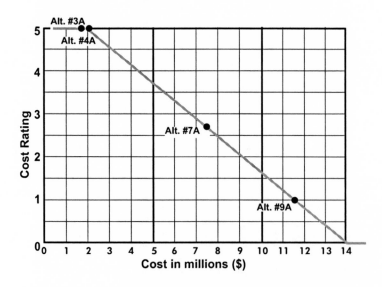

Exhibit 9.8a: Cost rating graph of STH 34 Bypass alternatives

Sensitivity Analysis

A summary evaluation of the project study alternatives is shown below. Using the average weighted rating for performance, acceptance and cost for each of the alternatives, the value indicators can be calculated with various weights of importance see Exhibit 9.9.

For example, the alternative 4A with weights of importance of P=2, A=1 and C=1,the value indicator is

$$\frac{2*2.4+1*2.7+1*5.0}{(2+1+1)} = 3.1$$

A sensitivity analysis, as shown in Exhibit 9.9, can be performed by assigning different weights of importance to the performance, acceptance or cost ratings.

Sensitivity Analysis

Alternatives			Do Nothing	Alt. 3A	Alt. 4A	Alt. 7A	Alt. 9A
Ratings Performance - P			2.8	2.1	2.4	4.2	4.5
Acceptance - A			3.6	2.6	2.7	3.7	4.0
Cost - C			5.0	5.0	5.0	2.8	1.0
	P	A	C		**Value Indicators**		
Weight of Importance	1	1	1	3.8	3.2	3.4	3.6	3.2
	2	1	1	3.6	3.0	(3.1)	3.7	3.5
	1	2	1	3.8	3.1	3.2	3.6	3.4
	1	1	2	4.1	3.7	3.8	3.4	2.6

$$\frac{2*2.4+1*2.7+1*5.0}{(2+1+1)} = 3.1$$

Exhibit 9.9: Value indicators for STH 34 Bypass alternatives

Alternative "Do Nothing" will not be considered as a viable alternative even though it has a better total rating. It has a fatal flaw in the individual ratings of performance and acceptance ratings. It is a good practice to drop any alternative that scored 0 or 1 in any individual rating unless it can be improved.

In general "Do Nothing" will score high on cost, ROW and constructability. If it does not fulfill the criteria that are vital to the improvement of the project, it should be dropped from further consideration. In some instances, the "Do Nothing" alternative will meet the minimum rating of 2 in all alternatives. In such cases it should be given due consideration.

Certain guidelines should be followed to rank the feasible alternatives. These guidelines will help narrow down the choices to make the final decision.

- The alternative should have a minimum average rating of 3.0 (Good) in performance, acceptance and cost.

- It should have an individual minimum rating of 2.0 on any criteria.

- Considering all three (performance, acceptance and cost), they should be ranked in the top 3 for various weights of importance.

Alternatives 3A and 4A have a low overall average weighted rating under performance and acceptance ratings. Alternative 9A has a low rating of 1.0 for the criteria *Time of Implementation*. Based on the above guidelines, the alternatives are ranked as follows:

1. Alternative 7A
2. Alternative 9A
3. Alternative 4A
4. Alternative 3A

Alternative 7A was the leading candidate. However, it was not the preferred alternative yet. The cost of this alternative was high. Therefore, the cost rating of this alternative was low. With these observations, the evaluation process was completed.

Conclusion

The Development Phase will explore alternatives further to make one of the leading alternatives a preferred alternative.

The Evaluation Phase will rank leading alternatives. The analysis matrices, performance and acceptance matrices will reveal the strengths and weaknesses of each alternative. During the Development Phase, the leading alternatives will be modified to enhance their chances of implementation.

Development Phase

Objective

The objective of this chapter is to improve or enhance the leading alternatives and select a preferred alternative, if it is feasible. Improvement is achieved by strengthening the performance and acceptance aspects of the alternative.

Introduction

In the Evaluation Phase, alternatives are ranked and rated. In the Development Phase, leading alternatives will be improved by a) combining ideas, b) enhancing good features of the alternatives or c) reducing or eliminating the negative features of the alternatives. Using the rating of the alternatives on each criteria, improvement can be made.

Process

Step 1: Review the ranked alternatives from the value indicator process as detailed in the last chapter

Step 2: List the low performance rating of the leading alternatives

Step 3: Explore whether these weaknesses can be eliminated

Step 4: Estimate the cost increase, if any

Step 5: Compare the features of all the alternatives, check whether any of these features can be incorporated in other alternatives

Step 6: Consult with specialists to improve the performance of the alternatives

Step 7: Calculate the function cost distribution of the leading alternative or alternatives to check whether a strong case can be presented

Step 8: Review the roadblocks which are non technical as indicated by the low acceptance ratings of the leading alternatives

Step 9: Develop a strategy to improve the acceptance of the leading alternatives

Step 10: Gather adequate information to develop a proposal to be presented (the detail of the proposal preparation is outlined in the next chapter)

Case Study – STH 34 Bypass Development of Alternative

The leading STH 34 Bypass alternative, 7A, was analyzed further for improvement. During the function analysis, it was observed that Alternative 7 had a value index of 3.75. Alternative 7 proposed three interchanges. The cost was $7.6 million. It was changed to Alternative 7A with partial interchanges at the two ends and no interchange at the middle intersection. The cost was reduced to $7.4 million.

The team questioned the need to build three bridges. The need for a grade separation on one end was tied to attracting new businesses. The other grade separation was projected to be needed if a new high school is built. These two major construction elements are meant to satisfy the design objectives *Attract Development* and *Sat-*

isfy Long-Term Needs. The team suggested that the provision should be made for future plans but there was no need to build it at the present time.

With this change the Alternative 7A was modified (see Exhibit 10.1). Alternative 7A Modified required a bridge across Maroons Road to access the hospital and a parallel frontage road for local access. Estimated cost of this alternative was $4.38 million.

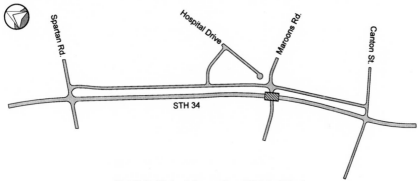

Exhibit 10.1: Alternative 7A Modified

The focus of the construction is to safely move traffic to meet the immediate needs. This is shown in the Technical Function Logic Diagram, Exhibits 10.2 and 10.2a. Also note that the design objective in Chapter 6 *Satisfy Long-Term Needs* is changed to *Accommodate Long-Term Needs.* Function *Satisfy Long-Term Needs* requires three bridges over the cross roads. But, function *Accommodate Long-Term Needs* requires the roadway profile to be designed to accommodate the future construction of two bridges over the STH 34 Bypass.

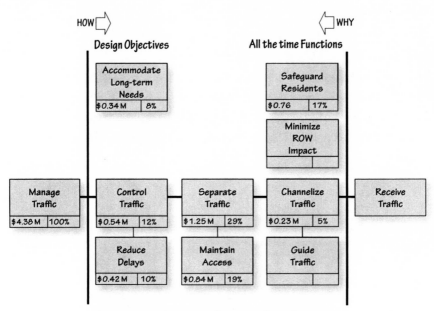

Exhibit 10.2: Technical Function Cost Diagram -
STH 34 Bypass VE Alternative 7A Modified

The cost of design objectives was reduced from 46 percent (As Given) to 17 percent (VE Alternative). The value index is

$$\text{Value Index} = \frac{\$4.38}{(\$0.54M + 1.25M + \$0.23M)} = 2.2$$

The desired value index, closer to 2.0 is achieved.

Customer function cost distribution for the Alternative 7A Modified is shown in Exhibits 10.3 and 10.3a. Value index for the Customer Function Logic Diagram for Alternative 7A Modified is calculated

$$\text{Value Index} = \frac{\$4.38}{(\$0.54M + \$2.24M)} = 1.56$$

Also note that the function cost distribution are not exactly the same for the two types of Function Logic Diagrams. They should always be in the same range.

	Element Cost	Control Traffic (Basic Function)	Separate Traffic	Channelize Traffic	Reduce Delays	Maintain Access	Safeguard Residents	Accommodate Long-Term Needs
		Critical Path			**Caused-By Functions**		**All-the-Time Functions**	**Design Objectives**
Structure	$712,800	$237,600	$237,600		$237,600			
Mainline	$240,000	$80,000			$80,000			$80,000
New Frontage Road	$2,550,000		$1,020,000			$510,000	$765,000	$255,000
MOT	$200,000				$100,000	$100,000		
Signals	$225,000	$225,000						
Intersection Improvements	$450,000			$225,000		$225,000		
Total - Maroons Road	**$4,377,800**	**$542,600**	**$1,257,600**	**$225,000**	**$417,600**	**$835,000**	**$765,000**	**$335,000**
Percentage	100%	12.39%	28.73%	5.14%	9.54%	19.07%	17.47%	7.65%

Exhibit 10.2a: Technical Function Cost

127

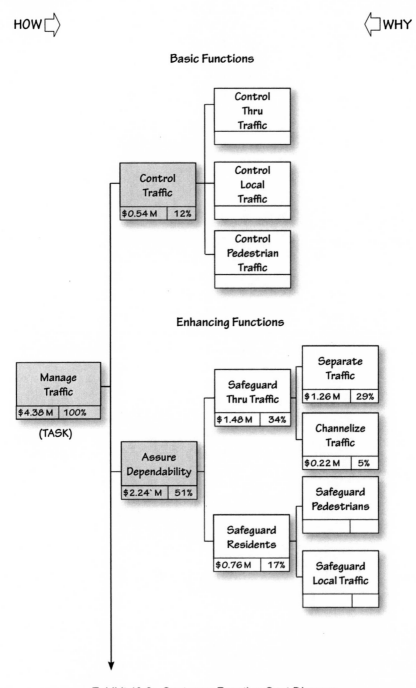

HOW⇨   ⇦WHY

Basic Functions

Enhancing Functions

Exhibit 10.3: Customer Function Cost Diagram

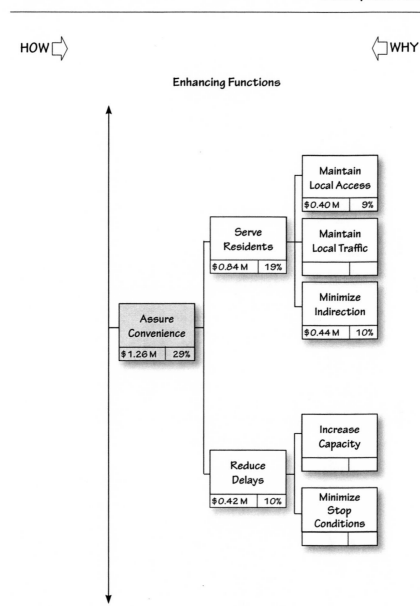

HOW⇨   ⇦WHY

Enhancing Functions

Exhibit 10.3: Customer Function Cost Diagram (Continued)

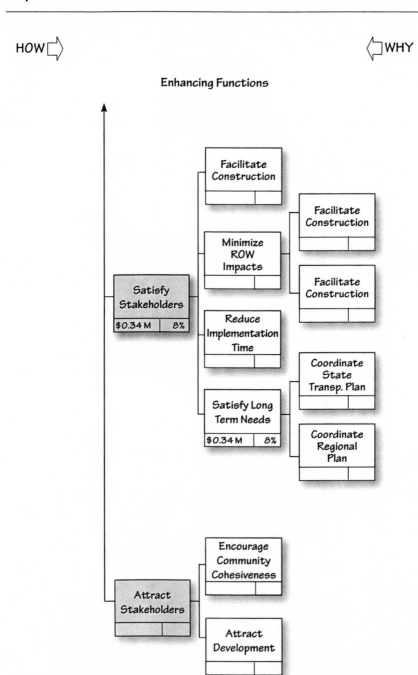

HOW ⇨   ⇦ WHY

Enhancing Functions

Exhibit 10.3: Customer Function Cost Diagram (Continued)

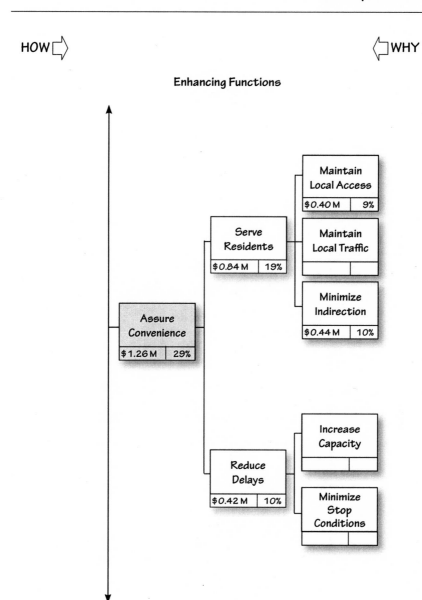

Exhibit 10.3: Customer Function Cost Diagram (Continued)

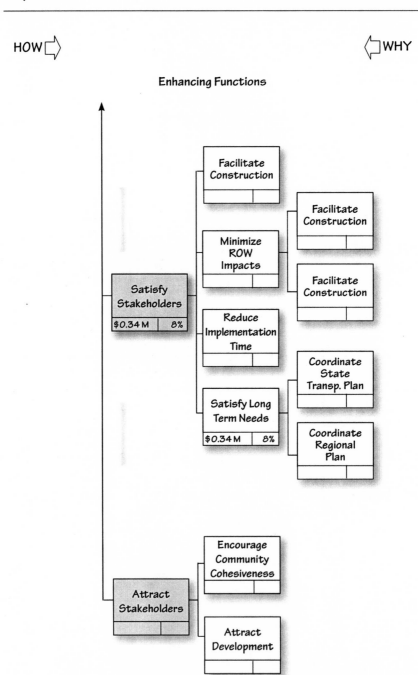

HOW⇨ ⇦WHY

Enhancing Functions

Exhibit 10.3: Customer Function Cost Diagram (Continued)

	Element Cost	Basic Function	Assure Dependability			Assure Convenience			Satisfy Stakeholders
		Control Traffic	Separate Traffic	Channelize Traffic	Safeguard Residents	Reduce Delays	Maintain Access	Minimize Indirection	Accommodate Long-Term Needs
Structure	$712,800	$237,600	$237,600			$237,600			
Mainline	$240,000	$80,000				$80,000			$80,000
New Frontage Road	$2,550,000		$1,020,000		$765,000		$255,000	$255,000	$255,000
MOT	$200,000					$100,000	$50,000	$50,000	
Signals	$225,000	$225,000							
Intersection Improvements	$450,000			$225,000			$95,000	$130,000	
Total - Maroons Road	$4,377,800	$542,600	$1,257,600	$225,000	$765,000	$417,600	$400,000	$435,000	$335,000
Percentage per function	100%	12.39%	28.73%	5.14%	17.47%	9.54%	9.14%	9.94%	7.65%
Percentage	100%	12.39%	51.34%				28.62%		7.65%

Exhibit 10.3a: Function Cost-Customer

131

The enhancement of value is tested by reviewing the Function Cost Diagram (Technical or Customer) or Ranking Matrices or both. In the Development Phase, check whether the stakeholders needs and desires are fulfilled within the constraints. In this case the needs of the thru traffic and local traffic were satisfied. However, the City's and County's long-term desires were not satisfied but were accommodated. This example is at the planning stage. If the VE study is during the design stage, more technical investigations and improvements would be applied.

Conclusion

The Development Phase is the link between selection and implementation. A thorough preparation of technical and financial details will set the stage to sell the alternative. The next step is to prepare the documents for an effective presentation.

Presentation Phase

Objective

The objective of this chapter is to convince the decision makers to accept the leading alternatives as a project solution.

Introduction

The implementation process begins with the Presentation Phase. The presentation may be to a single individual or group of four or more people. The media of presentation varies from a single document to a formal presentation. In all cases the presenter should know the audience. The strategy of selling the ideas varies with the two sides: the presenter and the audience.

Strategy

Three important steps can help the presenter get the message through; inform, instruct and influence. The message should always be simple yet powerful. In an informal manner the information can be conveyed to key players. Using the function logic approach is one way of preparing the materials for presentation:

- It should cover the basic information about the project and issues

- It should be dependable, error free, no contradictions and should support the recommendation

- It should be a convenient, easy to follow, clear message, with sketches to explain the presentation

- It should satisfy their expectations. In this case study, city, county and state officials had different expectations. The presentation should address all possible perceived problems, in addition to the real needs.

- It should be attractive and visual with attractive graphics. Drawings are better than mere numbers.

Essential Elements for Team Presentation

Understand audience

Before preparing the presentation, choose a media that suits the key audience. When presenting any new ideas or ideas that deviate from the audience's perspective, they may be viewed as a threat. People sometimes emotionally block out the positive side of the idea if it is not clear. Keeping this in mind, present your ideas in a simple and clear manner. Utilize visual media to get people's attention quickly.

Use the facilitative approach

The presentation should not have a tone of preaching nor of instructing. The presenter should draw the audience into participation. The presentation material should have a smooth flow, as if telling a story. If interrupted, take advantage of the question or comment to incorporate them into what you are trying to convey.

Maintain timing

Developing the sequence for how the information is presented is an art. Properly done, it produces good outcome. Short presentations may result in incomplete or lack of information. Longer presentation may lose the audience's interest in the subject. A 30 to 45 minute presentation is the suggested duration for a good presentation.

Ground Rules

At the conclusion of any presentation, the presenter should continue to influence the audience. It includes defending the presentation in a diplomatic way. During the question and response session, keep the following tips in mind:

1. **Listen first and respond later.** During the course of a discussion, be careful not to talk excessively. The audience wants to be heard. If you deny that privilege, he/she stops listening to you.

2. **Do not interrupt.** Don't interrupt a person when he/she is in the middle of their talk. It irritates the person. The way to win an argument is to avoid all appearances of argument.

3. **Don't create an argument.** Don't become belligerent or possessive, and don't let your voice get a tone of harsh contradiction.

4. **Inquire first and answer later.** During the first half of an argument or question, inquire first. Let the audience express his/her objections or opinions and fire off their heavy ammunition. Prepare your answers and respond.

5. **Repeat the person's objections briefly before answering.** Whenever a person objects or criticizes your position, calmly repeat what the person said. This gives the person an opportunity to listen to what they have just said. Let the person accept the fact that you understand their position.

6. **Concentrate on one issue at a time.** Focus on one key issue to win over. Give in to smaller unimportant objections. Repeat the responses that favor your position.

Case Study - STH 34 Bypass Development of a GRAND Illustration

Enhancement is achieved by visually showing the strength of the preferred alternative and the weakness of the other alternatives. Visualization has many parts: sketches, photos and other graphics. Special efforts should be taken to emphasize visualization. Alfred Benesch & Company has a elite visual graphics group, Group B, whose mission is to develop visual effects to help decision makers understand and appreciate the positive or negative features of any idea. Muthiah Kasi has developed a visual feature of the evaluation process. It is called GRAND (Graphic Analysis Diagram) illustration. This section describes the development of a GRAND illustration.

The process of developing a GRAND illustration is explained by using the Alternative 3A for performance rating. The performance rating is shown in Exhibit 11.1.

Excellent = 5 / Very Good = 4 / Good = 3 / Satisfactory = 2 / Poor = 1 / Unsatisfactory = 0	Weight of Importance	Do Nothing		Alt. 3A		Alt. 4A		Alt. 7A		Alt. 9A	
		Rating	Weighted Rating	Rating	Weighted Rating	Rating	Weighted Rating	Rating	Weighted Rating	Rating	Weighted Rating
Criteria	(1-10)	(1-5)		(1-5)		(1-5)		(1-5)		(1-5)	
1. Thru Traffic Safety	10	3.0	30.0	1.5	15.0	2.0	20.0	4.5	45.0	5.0	50.0
2. Local Traffic Safety	10	1.0	10.0	2.0	20.0	2.5	25.0	4.5	45.0	4.0	40.0
3. Pedestrian Safety	9	1.0	9.0	2.0	18.0	2.5	22.5	4.5	40.5	3.0	27.0
4. Thru Traffic Movement	8	4.0	32.0	1.0	8.0	2.0	16.0	4.5	36.0	5.0	40.0
5. Local Traffic Movement	7	2.0	14.0	2.0	14.0	3.0	21.0	4.5	31.5	4.5	31.5
6. Access to STH 34	6	5.0	30.0	3.5	21.0	2.5	15.0	3.0	18.0	3.0	18.0
7. Access from STH 34	7	5.0	35.0	3.5	24.5	2.5	17.5	3.0	21.0	3.5	24.5
Total Weighted Rating	57		160.0		120.5		137.0		237.0		231.0
Average Weighted Rating		2.81		2.11		2.40		4.16		4.05	

Exhibit 11.1: Performance Matrix

Step 1: Draw seven radiating lines representing the seven criteria. Make the angles between adjacent lines equal. For example, set a scale of 1 inch = 50 units. (The maximum weight of importance is 10. Maximum rating for excellent is 5. The maximum weighted rating will not exceed 5 * 10 =50.). Label each radiating line with one of the criteria (see Exhibit 11.2a).

Step 2: Develop the envelope for the excellent rating. If the weights of importance of all the criteria are 10, the envelope would be a symmetrical polygon. Since they are not the same the diagram will be an unsymmetrical polygon. For each criterion, mark a point equal to the product of the weight of importance and 5 for the excellent envelope. For the criterion "Access to STH 34" the length will be 6 * 5 = 30. Then join the points to develop an excellent envelope for Excellent (5) (see Exhibit 11.2b).

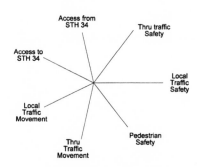

Exhibit 11.2a:

Step 1, Criteria lines with label

Exhibit 11.2b:

Step 2, Excellent criteria envelop

Step 3: Develop similar envelopes for Very Good (4) and Good (3) (see Exhibit 11.2c). The diagram shows the Excellent (5), Very Good (4) and Good (3) envelopes. The space within the Good (3) rating includes the Satisfactory (2) and Poor (1) envelopes. When the alternative is displayed graphically in this diagram, any white space within Good (3) envelope shows the weakness of the alternative. Ideally an excellent alternative will occupy the entire space within the Excellent (5) envelope.

Step 4: The next step is to plot the ratings of each alternative. For Alternative 3A, Access to STH 34 has a rating of 3.5 and a weighted rating of 21 (see Exhibit 11.1). Plot the point along Access STH 34 at a distance of 21. Use the same scale of 1 inch = 50. Similarly plot the points for each criterion. Exhibit 11.2d shows the rating of Alternative 3A in a graphical manner.

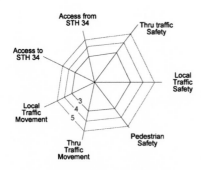

Exhibit 11.2c: Step 3,
Criteria envelop for other ratings

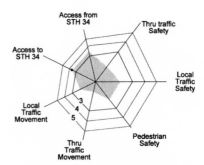

Exhibit 11.2d: Step 4, GRAND
illustration completed for
Alternative 3A.

Alternative 7A (237) and 9A (231) have nearly equal total ratings for performance (see Exhibits 11.3a and 11.3b). However, their distribution among the criteria is not similarly distributed. This is also true for acceptance rating (see Exhibits 11.4, 11.4a and 11.4b). The acceptance rating was not evenly distributed for Alternative 9A. This led to focusing on Alternative 7A.

As discussed in the Development Phase, Alternative 7A Modified does not require additional ROW since the bridges over Spartan Road and Canton Street are eliminated. The rating from the time of implementation is increased from 2.5 to 4.0 for Alternative 7A Modified. See GRAND illustration Exhibit 11.4a.

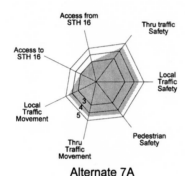

Alternate 7A

Exhibit 11.3a: Alternative 7A Performance Rating

Alternate 9A

Exhibit 11.3b: Alternative 9A Performance Rating

Excellent = 5 Very Good = 4 Good = 3 Satisfactory = 2 Poor = 1 Unsatisfactory = 0	Weight of Importance	Do Nothing		Alt. 3A		Alt. 4A		Alt. 7A		Alt. 9A	
		Rating	Weighted Rating	Rating	Weighted Rating	Rating	Weighted Rating	Rating	Weighted Rating	Rating	Weighted Rating
Criteria	(1-10)	(1-5)		(1-5)		(1-5)		(1-5)		(1-5)	
1. Compatibility w/Long Term	10	4.0	40.0	1.0	10.0	2.0	20.0	4.5	45.0	5.0	50.0
2. Compatibility w/Reg. Plan	9	4.0	36.0	1.0	9.0	2.0	18.0	4.0	36.0	5.0	45.0
3. Public Acceptability	9	0.0	0.0	3.0	27.0	3.0	27.0	4.5	40.5	5.0	45.0
4. Time of Implementation	8	5.0	40.0	4.0	32.0	3.0	24.0	2.5	20.0	1.0	8.0
5. Effect of Future Development	7	3.0	21.0	2.0	14.0	3.0	21.0	4.0	28.0	5.0	35.0
6. ROW Impact	6	5.0	30.0	4.5	27.0	3.5	21.0	3.5	21.0	1.0	6.0
7. Constructability	5	5.0	25.0	4.0	20.0	3.0	15.0	2.0	10.0	5.0	25.0
Total Weighted Rating	54		192.0		139.0		146.0		200.5		214.0
Average Weighted Rating		3.56		2.57		2.70		3.71		3.96	

Exhibit 11.4: Acceptance Matrix

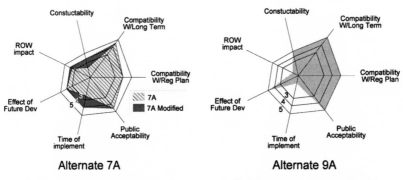

Alternate 7A

Exhibit 11.4a: Alternative 7A &
7A Modified
Acceptance Rating

Alternate 9A

Exhibit 11.4b: Alternative 9A
Acceptance Rating

This process helps people understand their needs and match them with the action. It should be the policy to spend money for today and make provisions for tomorrow. VE helps you avoid two scenarios: spending money for uncertain, future needs and approving a design that will hinder future planned construction.

Proposals and Design Suggestions

Part of selling the preferred alternatives is to present the following steps:

Step 1: Describe existing condition in great detail

Step 2: Describe the "As Given" or "As Designed" solution

Step 3: Introduce the VE alternatives

Step 4: Compare the "As Given" with the "VE Alternatives"

- Cost and Life Cycle Cost, if appropriate

- Advantages

- Disadvantages or limitations

In each case develop sketches and other support materials. The presentation will be introduced as a proposal or design suggestion.

The VE presentation may be a complete alternative to the "As Given" or it could be individual ideas. When the alternative can be supported by cost, design features and clear advantage over the "As Given," it is classified as a "Proposal." When they lack part of the above mentioned features, it is classified as a "Design Suggestion" (see Exhibit 11.5). Proposals, if the clients agree, will become solutions. Design Suggestions, if accepted, will be included by the design team in their design as a VE solution.

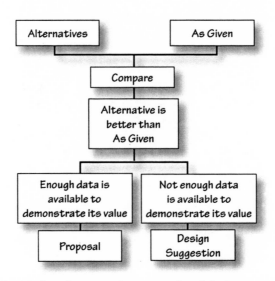

Exhibit 11.5: Development of a proposal and design suggestion

Outline of a Proposal

Existing Conditions

Write a short description of the existing conditions. This should include a discussion of the issue(s) with the existing conditions that have led to the decision to repair, rehabilitate or reconstruct the item.

Include a photograph, sketch or plan of the existing condition.

As Given

Write a short description of the As Given. This should include a discussion of the attributes of the As Given Solution that resolves the problems with the existing conditions.

Include a sketch of the As Given, a cost estimate that details the major pay items and/or materials that resolve the problems with the existing conditions and the functions it satisfied.

Value Engineering Alternative(s)

Write a short description of the Value Engineering Alternative(s). This should include a discussion of how it resolves the issues with the existing conditions as well as, or better than, the As Given. List the functions that are satisfied by Value Engineering Alternatives.

Include a sketch of the Value Engineering Alternative(s), a cost estimate that details the major pay items and/or materials and the functions it satisfied.

Advantages/Limitation of Value Engineering Alternative(s) vs. As Given

Write a short description of the advantages and limitations of both. This should cover the advantages and limitations from both a performance and acceptance perspective.

Recommended Solution

Write a short summation of the reasons why we are recommending a VE alternative.

When the VE proposals are presented, the owner has four choices:

1. **Accept:** It is what every value engineer hopes for.

2. **Reject:** If it is rejected, the team should find out why. If it is a constraint, check whether the team was told of this constraint. If not, investigate the reasons for not knowing the constraint. Change your strategy for future studies to flush out these constraints. Make sure that the information the team gathered is sufficient.

3. **Needs further study:** This is mostly an internal issue of the owners. They need to make up their mind on outstanding issues concerning the needs, desires and constraints of the owner, user and stakeholders. Sometimes it may be techni-

cal issues that they are not comfortable with. In our previous study on the truss shortening, it was never done before. The owner wanted the consultant to answer all concerns.

4. **Consider for future application:** An idea may be good, however, it may be rejected due to project specific constraints such as schedule or previous commitments. The owner may accept this as a good idea for future applications. In such cases, the designers will make a note of it to use this idea in the future projects.

The owner/decision maker will select one of the four choices for each proposal.

Case Study - STH 34 Bypass Proposal

Exhibit 11.6 is an example of the STH 34 Bypass proposal.

STH 34 Bypass Improvements	Proposal No 1 1 of 4

Existing Condition

STH 34 in Coldwater, Wisconsin was originally intended to be a by-pass around the City. However, developments over 20 years made this bypass into a local roadway. This resulted in conflicts between the thru traffic and the local traffic.

The accident rate was much higher than the statewide average for this type of highway. The city, county and state officials want a solution that will improve the safety condition considerably while the convenience of the through and local traffic are not compromised.

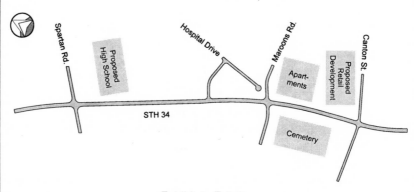

Exhibit A: Existing

As Given

Construct three bridges along the existing STH 34 Bypass. The bridges will carry STH 34 Bypass over Canton Street, Maroons Road and Spartan Road. The three crossings will be signalized below the bridges. The three crossings will have a tight diamond interchanges to access the town (see Exhibit B).

Exhibit 11.6: STH 34 Bypass proposal

STH 34 Bypass Improvements	Proposal No 1 2 of 4

Exhibit B: Alternative 7 (As Given)

Cost of this improvement is $7.6 million

Advantages:
- It maintains the thru traffic speed
- It improves the safety of the thru and local traffic
- It provides better access to the community
- It satisfies the long term plan of the region
- If the school is built, it offers a convenient and safer crossing for the future high school students
- It encourages and attracts future commercial developments near Canton Street

Limitations:
- It costs more than the program cost
- It assumes the school and development will utilize the added crossings. If either one doesn't materialize, the investment is a mismatch.
- The three exits and entrance ramps require additional ROW
- The three exits and entrance ramps impede the thru traffic

Exhibit 11.6: STH 34 Bypass proposal (continued)

STH 34 Bypass Improvements	Proposal No 1 3 of 4

VE ALTERNATIVE 7A MODIFIED

Alternative 7 was revised to have three bridges with access at the beginning and the end of the city limits. Construct a frontage road adjacent to the existing STH 34 Bypass. This alternative is shown in Exhibit C as Alternative 7A. It was later modified to limit to one bridge across Maroons Road and a parallel frontage road was built to serve local traffic. The access to STH 34 Bypass was at Canton Street and Spartan Road. The VE alternative, Alternative 7A Modified is shown in Exhibit D.

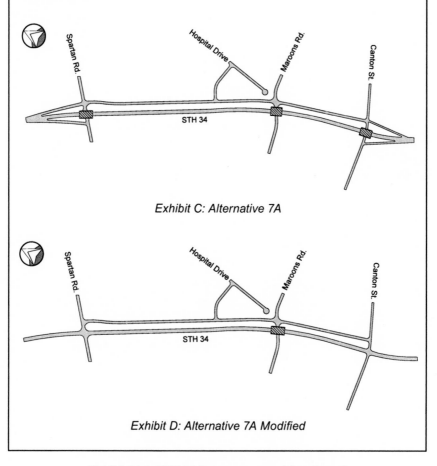

Exhibit C: Alternative 7A

Exhibit D: Alternative 7A Modified

Exhibit 11.6: STH 34 Bypass proposal (continued)

STH 34 Bypass Improvements	Proposal No 1 4 of 4

The cost of Alternative 7A Modified is $4.38 million.

Advantages:
- It maintains the thru traffic speed
- It improves the safety of the thru and local traffic
- It provides better access to the community
- It accommodates the long term plan of the region
- If the school is built, a new bridge can be built
- If commercial development near Canton Street is a possibility, a new bridge can be built
- It meets the program cost

Limitations:
- It slows the traffic at Spartan Road and at Canton Street crossings.

RECOMMENDATION:
The VE team recommends the adoption of Alternative 7A Modified. The immediate need is to cross along Maroons Road and to access the hospital area. The crossing traffic is expected to increase along Spartans Road and Canton Street. The bridges can be easily added on an as needed basis in the future. This alternative meets the program cost. It satisfies the needs and desires of the project without violating any project constraints.

Item	First Cost		Maintenance & Operation Cost		Cost Difference
	As Designed	Proposed VE Alternative	As Designed	Proposed VE Alternative	
Alternative 7 Alternative 7A Alternative 7A mod.	$7.61 $7.51 $4.38				

Accepted	_____	Needs Further Study	_____
Rejected	_____	Consider for Future Application	_____

This recommendation was later adopted. The bridge and the frontage road were built 15 years ago. Since then the residents voted down the location of the high school in the vicinity of the STH 34 Bypass. The need for the bridge over Canton Street didn't materialize.

Exhibit 11.6: STH 34 Bypass proposal (continued)

Conclusion

In this chapter we learned the importance of documentation and communication. Communication includes written, visual and oral. Oral and visual presentations can encourage the decision makers to read what is presented. This will enhance the chance of accepting the recommended solution.

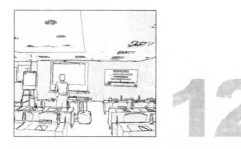

Organizing a VE Study

Objective

The objective of this chapter is to show the critical steps to a successful VE study.

Introduction

In the previous chapters the basic principles of VE were described. Every value engineering study is, in some respects, unique but there are some steps that need to be taken prior to the actual study that will make the organization of the workshop and the study smoother. The VE team leader has to understand these steps, and be patient and diplomatic to strategize and implement them.

In general, the steps to be taken by the team leader are:

- Team Selection
- Subject Selection

Team Selection

Team size may vary, but a team of five is preferable. Having less than four seriously limits the amount and variety of creative input, and more than six tends to be unwieldy and time consuming.

Less than	3	is too small
	4	stifles interaction
	5	is ideal
	6	is slightly less than ideal
		(A larger number, by virtue of their size, may have problems of interacting meaningfully)
	7	encourages multiple conversations and discourages team spirit
	8	means splitting into multiple teams
Greater than	9	is a group conducting a meeting not a workshop

If more than 10 members want to participate, divide the team into two or more teams of five.

Selection of team members is critical in organizing a study. People with expertise in different disciplines should be included as consultants, but not members, of the VE team on an as needed basis. Apart from the technical considerations, it is advisable to select engineers who do not think alike. It is healthy to have a mixture of talents and temperaments that will generate questions and even arguments. Resolution of questions raised by team members ensures better solutions. This is the only way a good study can be conducted.

Team members should have a mix of talents:

- Logical/methodical
- Creative
- Optimistic

- Clerical
- Curious
- Emotional

It requires mixed skills to be a good team member:

- Philosopher
- Dreamer
- Coordinator
- Secretary
- Doer
- Decision maker

The team leader should participate in the selection of the team members for any given study. It is desirable that, in addition to the team leader, one team member participates in all the studies in an organization. The familiarity and continuity of at least one additional team member helps. The team leader can rely on this person to conduct the study in case of an emergency. The team leader should be a person who is well versed in all phases of the project, but not necessarily an expert in any given field. The team leader can control and guide a study in the proper path by being knowledgeable and assertive, yet tactful and friendly. The team leader should be able to think ahead and present ideas clearly and with perseverance. A team leader who steers a study should not have the following weaknesses. The leader should work to eliminate the weaknesses or have someone help the leaders during or prior to the workshop to avoid situations.

- Weak leadership (afraid to take a position)
- Not focused

- Too rigid
- Failure to recognize strength/weakness of team members
- Unprepared
- Afraid to decide on direction

The team should be represented by appropriate disciplines so that the results are not favored by one discipline over another. Depending upon the type of project the team members are usually selected from the following disciplines:

- Traffic
- Structural
- Drainage
- Geometrics
- Construction
- Cost
- Signage
- Highway
- Scheduling

Subject Selection

In most instances, the team is given the opportunity to select the elements to be studied. In making the selection, the team must keep in mind that while value engineering analysis results in cost savings and improvement in any element, it is time consuming and thus costly to apply to all elements of a design project. A general rule of thumb is that a study should not be undertaken unless the following applies:

Expected first cost savings > 10 x cost of study

Therefore, elements having a low first cost should not be considered unless their need for high acceptance is perceived to be great. For example, all safety functions require high acceptance, therefore any element whose basic function is "Safeguard User" should be studied if the team perceives the performance of the element to be less than acceptable.

In addition, the following criteria are guides in selecting an element for study:

- If first cost exceeds initial estimated cost or budget
- If the element has an obviously unwanted costly function
- If the element is difficult to construct
- If the element has too many non-standard details
- If the element is obviously difficult to maintain

There are three other factors used to identify the elements that are worthwhile subjects:

- Pareto's Law of Distribution
- Probability of Acceptance of Change
- Probability of High Maintenance Cost

Pareto's Law of Distribution

In the nineteenth century, an Italian economist (Pareto) developed a curve (Exhibit 12.1) known as Pareto's Law of Distribution. This curve applies whenever a significant number of elements are involved. It states that in any area, a small number of elements will represent the major cost.

Exhibit 12.1a shows a modified version of Pareto's Law of Distribution for a stringer bridge. Since this bridge was located in a poor soil area, where settlement was critical, the foundation cost was high.

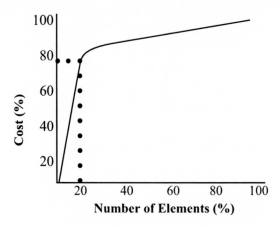

Exhibit 12.1: Pareto's Law Of Distribution

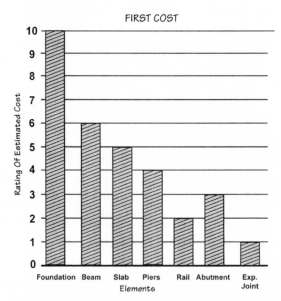

Exhibit 12.1a: Modified version of Pareto's Law of Distribution for a stringer bridge

Probability of Acceptance of Change

Using their past experience, the team should rate the elements for the probability of acceptance of change. To be successful, the change should have a high probability of being accepted by the project designer, management and the client. It is very important to note that this does not mean that elements with low probability of acceptance of change should not be studied. It merely guides the team in setting the priority for which element to study. Exhibit 12.2 shows the probability of acceptance of change for the bridge elements.

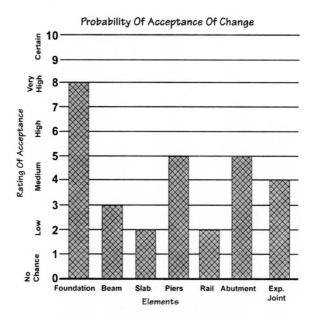

Exhibit 12.2: Probability of acceptance of change for the bridge elements

Probability of High Maintenance Cost

Project engineers should keep track of the cost of maintenance of various elements of structures designed by their firm. Value engineers are responsible for identifying high life cycle cost elements and insisting on value engineering studies to reduce maintenance cost. Knowledgeable persons, particularly members of the owner's maintenance force, should be consulted. Exhibit 12.3 shows the probability that the bridge elements will require maintenance.

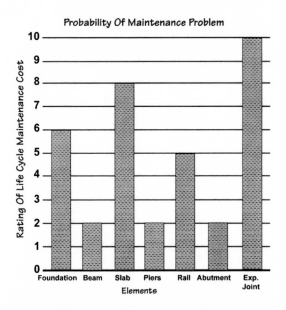

Exhibit 12.3: Probability that the bridge elements will require maintenance

Weights of Importance

The team should determine the weight of importance of each of the factors discussed. For the bridge, the following weights were used:

First Cost	6
Acceptance of Change	10
Maintenance	8

These weights of importance are multiplied by the rating received for each of the bridge elements to obtain a score. The foundation received a rating of 10 for the cost and a score of 60 (*i.e.*, 10 x 6). The scores for the three factors are now added to obtain the weighted total. Finally, the top weighted total is assigned 10 and the others are a ratio of this number (82/188 x 10 = 4.4). These relative ratings are shown in chart form (Exhibit 12.4) and as a graph for comparison (Exhibit 12.5). Selection of elements to be studied will be made on these ratings.

Relative Rating Of Elements					
	First Cost	Acceptance	Maintenance	Total (Weighted) Ratings	Relative Rating
Weights of Importance	**6**	**10**	**8**		
Foundation	60	80	48	188	10.0
Beam	36	30	16	82	4.4
Slab	30	20	80	130	6.9
Piers	24	50	16	90	4.8
Rail	12	20	48	80	4.3
Abutment	18	50	16	84	4.5
Expansion Joint	6	40	80	126	6.7

Exhibit 12.4: Relative rating of elements

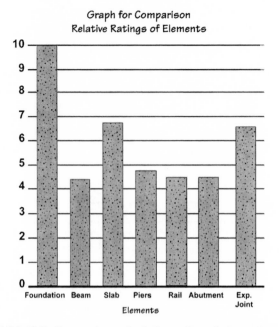

Exhibit 12.5: Comparison of relative rating of elements

For the bridge, the expansion joint is not an expensive item, ranking lowest in the cost consideration. However, it ranks third in importance in the study, as shown in the composite diagram, for two reasons: a) Expansion joints, in general, cost more to maintain or repair. b) The owner gives the joint a higher priority for maintenance than other elements. Exhibits 12.1b thru 12.5 vary for each project and depend on the following: structure, owner, time and location

Conclusion

The team should develop figures similar to Exhibits 12.1b - 12.5 before selecting the element(s) to be studied.

Conclusion

Introduction

This chapter concludes the book with certain key points to remember as a value engineer. First is to understand what value is and how we can achieve it. Second is the timing of performing a VE study. If a pre-bid VE study was not performed, keep in mind that a Value Engineering Change Proposal (VECP) should be considered. The third key point is how to work as a team in a formal workshop format.

Value is in the Eye of the Holder

The reasons or purpose of a project of similar type cannot be assumed to be the same all the time. The VE team should not be the judge to decide what is value. The following demonstrate how the concept of value can vary.

Larry Miles termed the VE process as "determining the possible." VE does not solve any problem. Instead, it shows the possibilities. It is up to the decision makers to determine what value it is they desire. Years ago, I conducted a series of VE studies on tunnels (Exhibit 13.1). In the United States, the practice is to have a raised sidewalk for maintenance staff and pedestrian emergency access. The increased cost of a raised sidewalk, as opposed to at grade sidewalk, was about $850,000. This is due to the enlarged tunnel cross section to give proper headroom for the pedestrians. The owners justified the cost for safety reasons. When I conducted a tunnel study in Taiwan, the thinking was totally different. They preferred an at-grade sidewalk with a curb for safety reasons. Their reasoning is that in case of accident, the occupants of the car may be injured. They won't be able to climb up the steps to the walk way. In addition, the access to the walkway is spaced 100' apart. By having it at grade they can crawl, roll over or step into it and avoid the traffic. The value perception, with respect to safety, is the opposite for the same type of structure. It is during the Information Phase, the team should explore, understand and agree on the definition of value.

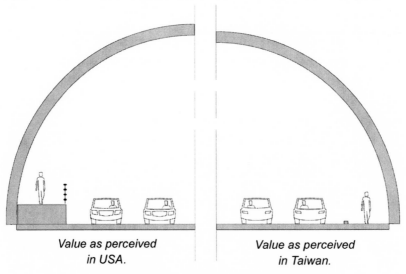

*Value as perceived
in USA.*

*Value as perceived
in Taiwan.*

Exhibit 13.1: VE studies on tunnels

Timing of VE study

For maximum benefit, VE should be initiated as early as possible in the project development process. This facilitates the implementation of VE recommendations without delaying the progress of the project. In such cases, significant rework of completed designs can be avoided. Roadblocks to a VE study can occur if it is too early to do a VE study since design hasn't started or too late to do a VE study since there is not enough time to make changes. Neither observation is valid. The strategy of selecting elements to study, developing the cost of the elements and the type of recommendations vary. Value engineering can be done at any stage. The strategy to conduct and implement the study will be different.

Value Engineering Change Proposal

Value Engineering Change Proposal, widely known as VECP, is the last chance an owner has to lower the cost and improve the value of the project. VECP also gives the contractor an incentive to seek ways and means to increase his profit without lowering the value of the project. A successful VECP benefits both the owner and contractor.

There is a perception that "If the designer did his job right in the first place, there is no need for VECP." This belief is false. The following reasons will justify need for VECP:

1. There is always a long time gap between the design and construction phases, and the market conditions may change dramatically during this period.

2. During construction, field conditions are more precise.

3. Constraints that exist during design may not exist during construction. An existing building that was close to the construction site during design phase might be demolished before the construction of an adjacent roadway.

4. Designers assume a method of operation and staging of construction suitable for all qualified bidders. The successful bidder, because of their unique expertise or location to the site, may be in a position to select a different or familiar construction for less cost.

A VECP would not be considered a weakness of the designer, but should be considered an opportunity for a positive team effort between designer and contractor.

A VECP should follow the following steps:

Step 1. Documentation

- Present comparison between design and VECP, detailing advantages and disadvantages

- Itemize changes to the contract

- Submit a cost estimate for both design and VECP including the cost of development and implementation of the redesign by the contractor

Step 2. Submission. The contractor should submit the VECP to the resident engineer with a copy to the owner

Step 3. Acceptance. The contractor should set a time limit for a response. Beyond this limit, the contractor has the option to withdraw the VECP proposal. The owner should have the right to accept or reject the proposal.

Step 4. Sharing. The contract will detail the method by which the contract price would be adjusted if the proposal is accepted.

Probability of Success - VECP

No matter how good a VECP is, its success depends mainly upon how receptive the designer is toward the changes, and how much the owner wants to participate in the technical aspect of the project. Without this receptivity and participation, the contractor may not want to risk time, money and the possibility of antagonizing the designer. In such a situation, either the contractor will not propose a VECP or will not pressure the owner or the designer to accept a suitable VECP.

Understanding Team Behavior

A formal VE study is normally a 40-hour workshop with five to seven members. During the later part of the Information Phase (analysis phase), people usually get frustrated when they are struggling to pinpoint the reason for the project elements. Team members do not have the patience to study the reasons for the project improvement. They are more anxious to come up with ideas which they, in most cases, assume as good solutions. Patience and team chemistry are important to the success of a good study. Each study should have a team leader and an anchor person. The team leader should monitor the mood of the participants and guide them through the study. The anchor person anticipates the needs of the project and collects information needed to create continuous flow of information, decisions and discussions.

Based on my years of observation of VE studies, I noticed a similar pattern of each team's reaction to a typical five day VE study, see Exhibit 13.2.

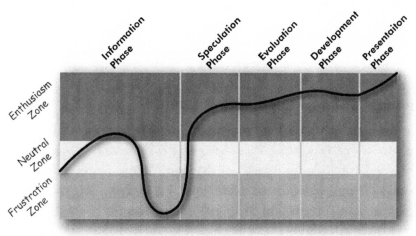

Exhibit 13.2: Graph of VE participants response during a typical 40-hour VE

During the Information Phase, a neutral mood changes to an enthusiastic mood as they learn more about the project and the opportunity to create new ways of doing it. When they are prohibited from being creative and are asked to define functions in two words with a How-Why logic they get frustrated. The frustration bottoms out as the Function Logic Diagram is completed. At this point, as they understand the task or higher order basic function, the enthusiasm increases. The team continues to be enthusiastic during the Speculation Phase. They remain in this mood during Evaluation Phase. As they prepare for presentation, they feel more confident and feel a sense of accomplishment. This pattern can be observed in most successful VE workshops. When this pattern changes, the study may not be going smoothly. The team leader should monitor for such signs and be prepared to make proper and timely course corrections if needed.

Micro VE studies

For small studies, the team's knowledge of the project, and their understanding of the VE process will help achieve a shorter time period. The same team members should perform a series of VE studies so that they can practice developing functions, structuring the Function Logic Diagram, listing ideas and selecting a solution. By performing a series of VE studies, the team will learn to understand each other and become very efficient in performing studies.

Concept of Value Engineering

The ultimate goal of a VE study is to carefully transform the needs and desires to functions, then speculate on ideas for all functions and develop a solution that scores high on performance with a reasonable acceptance and cost. Exhibit 13.4 shows the diagram of the VE process as described in the previous chapters. At the end, all VE efforts intend to satisfy the owners, users and stakeholders. The VE team should keep the following three points in mind when searching for value.

1. Every action is required or desired by someone (Stakeholders)

2. Every action has a reason or purpose (Function)

3. The cost of each action must be justified within the limit of the constraint (Function Cost)

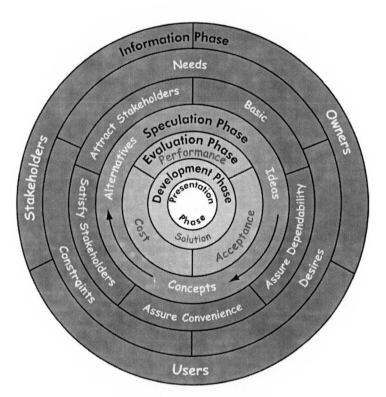

Exhibit 13.3: VE Job Plan

Exhibit 13.4 shows the process from needs and desires of stake-holders to the project solution using the VE Job Plan (Exhibit 13.3). Exhibit 13.4 flow chart is the summary of the Benesch VE process discussed in this book.

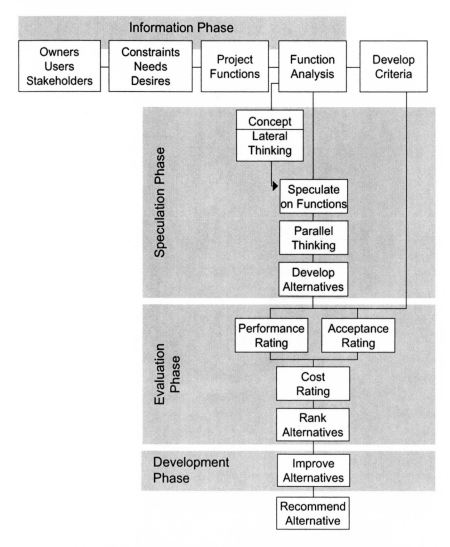

Exhibit 13.4: Flow chart summarizing the Benesch VE process

Conclusion

Value engineering is a structured, team oriented concept. As we participate in more and more VE studies, we understand the concept better. As we gain experience, we will think in terms of functions all the time. Whether you buy a car or house or a suit you will notice that your mind will start asking the question "What does it do?" You start looking at various options and start judging its performance, acceptance and cost. When you are ready to answer the question "Why?" without hesitation, you understand the meaning of the term, "value." When one reaches this stage they have truly become a value engineer.

Glossary

Acceptance: The degree of a favorable response from the stakeholders.

Alternative: A substitution for another solution. It is usually the result of lateral thinking.

Basic Function: An action that is deemed necessary to fulfill the task.

Concept: An abstract or generalized thought. In value engineering, the concept may be focused on one of the four enhancing functions, a concept stressing dependability or convenience, or satisfying or attracting a stakeholder.

Constraints: Are defined by the stakeholders. They also include legal requirements, standards of the owner, physical restrictions of the site or commitments to stakeholders by the owner. These stakeholder constraints will be considered in developing the project constraints.

Customer–Oriented Function Logic Diagram: A diagram that displays the customer needs and desires. See also Function Logic Diagram. This concept was developed by

Thomas Snodgrass and Ted Fowler.

Design Suggestion: A suggested design feature which does not have all the design details or cost and may be considered in the Design Phase.

Desires: These are the expectations of the stakeholders that are on a wish list. These will be considered if they are affordable.

Elemental Cost: Allocating cost to various elements of a structure or a project (*i.e.* pier, pavement, shoulder).

Enhancing Functions: See Supporting Functions.

FAST: See Function Logic Diagram.

Function Logic Diagram: A logical display of the project functions in a How-Why format used clearly defines the needs and purpose of the project. There are two types of function logic diagrams: Technical Function Logic Diagram and Customer-Oriented Function Logic Diagram. The original and official name is FAST diagram.

Functions: The needs and desires of the stakeholders as seen from the project point of view and expressed in two words; an active verb and descriptive noun.

Function Cost: Allocation of cost of components to a function.

GRAND: (Graphic Analysis Diagram) A graphic display of the evaluation of alternatives. This concept was developed by Muthiah Kasi.

Idea: A descriptive representation of a concept. In value engineering, an idea is a descriptive representation of a function.

Indirection: Not following the shortest way to a destination.

Job Plan: A structured process with defined phases. Each phase has guidelines that should be followed.

Lateral Thinking: Seeks to drastically change the existing condition. Lateral thinking is concerned with the perception part of thinking. This will result in the creation of out-of-the-box concepts.

Mismatch: When the need of the function is low and the cost is high. This concept was developed by Thomas Cook.

Needs: The expectations of the stakeholders that must be fulfilled by the project, if the project constraints are not violated or other stakeholder's needs are not in conflict.

Owner: a person/organization that funds the project or is responsible for the project.

Parallel Thinking: A cooperative and coordinated thought of each concept or lateral thinking concepts.

Options: An item or action that is offered in addition to, or in place of, the standard. Options usually are the result of parallel thinking.

Performance: A measurement of how well an alternative meets the functions of the project.

Project: A group of actions or tasks that will satisfy one common purpose. A bridge or road design/construction may be a task or project. Production of coffee mugs may be a task or project.

Proposal: A recommended alternative describing the design with support details of design features, its advantages, limitations and cost.

SCREAM: A speculation technique which suggests changes to the concept by Substitute, Combine, Rearrange, Eliminate, Adopt or Modify an idea.

Screening Graph: A graphical display of the initial screening of ideas developed in the Speculation Phase. This concept was developed by Muthiah Kasi.

Stakeholder: Anyone who is financially or environmentally affected by the project. Owners and users are also stakeholders.

Supporting Functions: Also known as Enhancing Functions. These functions are essential to the enhancement and support of the basic functions.

Task: A function that satisfies the overall needs and purpose of the project.

Team Leader: A Certified Value Specialist who leads the VE study.

Technical Function Logic Diagram: A form of Function Logic Diagram that describes the designer's point of view. See also Function Logic Diagram.

Trade Cost: Cost allocated to various materials or actions of a project (*i.e.* concrete, steel, excavation).

TQE®: A registered trademark of Alfred Benesch & Company which combines various levels of value management techniques (Total Quality Engineering). Developed by Muthiah Kasi and Michael N. Goodkind.

User: One who actively or physically uses the project or maintains the project.

Value: When the alternative reliably performs all the needed functions, is acceptable to the stakeholders and comes at a reasonable cost to the client.

Value Engineering: A systematic team approach used to analyze and creatively enhance the value of a project or product.

Value Engineering Change Proposal (VECP): A contractual clause that is available to the contractor to propose a value engineering change to the owner for consideration. If accepted the owner and contractor will share the savings.

Value Index: There are two definitions based on the type of Function Logic Diagram. For Technical Function Logic Diagram, the value index is the ratio of the total cost of the project to the cost of the critical path functions. For the Customer-

Oriented Function Logic Diagram, the value index is the ratio of the total cost of the project to the cost of *Basic* and *Assure Dependability* functions. (This value index was set by Muthiah Kasi.)

Value Indicator: The average weighted rating summary of performance, acceptance and cost ratings. This detailed approach was developed by Muthiah Kasi and Michael N. Goodkind.

References

1.   Miles, Larry D., Techniques of Value Analysis, 3rd Edition, (McGraw Hill Book Company, 1989)

2.   Snodgrass, Thomas J. and Kasi, Muthiah, Function Analysis–The Stepping Stone to Good Value (University of Wisconsin, 1983)

3.   ASTM International, "ASTM Standard E 2013 – Standard Practice for Constructing FAST Diagrams and Performing Function Analysis During Value analysis Study," (2006)

4.   ASTM International, "ASTM Standard E 1699 – Standard Practice for Performing Value Analysis of Buildings and Building Systems," (2005)

5.   ASTM International, "ASTM Standard E 2103 – Standard Classification for Bridge Elements and Related Approach Work," (2006)

6.   De Bono, Edward, Serious Creativity (Harper Business, 1992)

7.   SAVE International, "Monograph Function: Definition and Analysis," (1998)

Index

LaVergne, TN USA
20 August 2009
155248LV00002B/5/P